I0015431

M5Stack Electronic Blueprints

A practical approach for building interactive electronic
controllers and IoT devices

Dr. Don Wilcher

BIRMINGHAM—MUMBAI

M5Stack Electronic Blueprints

Copyright © 2023 Packt Publishing

All rights reserved. No part of this book may be reproduced, stored in a retrieval system, or transmitted in any form or by any means, without the prior written permission of the publisher, except in the case of brief quotations embedded in critical articles or reviews.

Every effort has been made in the preparation of this book to ensure the accuracy of the information presented. However, the information contained in this book is sold without warranty, either express or implied. Neither the author(s), nor Packt Publishing or its dealers and distributors, will be held liable for any damages caused or alleged to have been caused directly or indirectly by this book.

Packt Publishing has endeavored to provide trademark information about all of the companies and products mentioned in this book by the appropriate use of capitals. However, Packt Publishing cannot guarantee the accuracy of this information.

Group Product Manager: Rahul Nair
Publishing Product Manager: Surbhi Suman
Senior Editors: Divya Vijayan and Athikho Sapuni Rishana
Technical Editor: Shruthi Shetty
Copy Editor: Safis Editing
Project Coordinator: Ashwin Kharwa
Proofreader: Safis Editing
Indexer: Subalakshmi Govindhan
Production Designer: Ponraj Dhandapani
Marketing Coordinator: Gaurav Christian
Senior Marketing Coordinator: Nimisha Dua

First published: March 2023

Production reference: 1180123

Published by Packt Publishing Ltd.
Livery Place
35 Livery Street
Birmingham
B3 2PB, UK.

ISBN 978-1-80323-030-6

www.packt.com

To my children Tiana, D'Vonn, and D'Mar, for allowing me to see them grow into wonderful young adults and for supporting my work as an electrical engineer, educator, scholar, researcher, and author. To Juwon, for asking critical questions about the M5Stack Core projects demonstrated in my lab. To my granddaughter Winter, for allowing me to play and learn through her eyes as a child. To Larita and Matthew, thank you for embracing me as your son. To my natural parents, thank you for providing a strong work ethic and an educational mindset. To my wife, Dr. Mattalene Wilcher, for being my loving partner throughout our joint life journey.

– Dr. Don Wilcher

Foreword

Don Wilcher has succeeded again with this new book about using the economical M5Stack Core microcontroller. This remarkable controller is designed to interface with a variety of circuits simply by stacking them atop the compact 5 cm x 5 cm M5Stack Core package. This and its block coding capability greatly simplify the design and implementation of compact controller packages for use in robotics, automation, wireless controllers, scientific applications, and a wide variety of other projects.

If you want to get up to speed with integrating one of the latest microcontrollers with a variety of external devices and circuits, this book is for you. Don is a hands-on engineer with outstanding communication skills. Moreover, he has built, programmed, and tested the systems he describes in this book. I was especially interested in *Chapter 2*'s description of interfacing an M5Stack Core microcontroller with a tri-color LED, and there's much, much more in this remarkable book.

Don Wilcher is a master electronics communicator, and I am pleased to recommend this book to hobbyists, educators, companies, and anyone interested in designing, assembling, and programming an M5Stack Core microcontroller.

Forrest M. Mims II

www.forrestmims.org

"One of the 50 best brains in science." Discover Magazine

Contributors

About the author

Dr. Don Wilcher is an electrical engineer, educator, maker, scholar, researcher, and entrepreneur experienced in mechatronics, embedded control systems, and electrical-electronics technologies. He has worked in a variety of product, systems engineering, management, and supervisory roles in the consumer, manufacturing, and automotive industries. He has extensive experience in designing and analyzing electromechanical and electronic control systems and prototyping wearable devices. Besides his industrial experience, Dr. Don Wilcher has reviewed the curriculum and industry standards for engineering, STEM, career, and technical education programs. He has redesigned technical courses in manufacturing, robotics, and mechatronics, and developed instructional materials for such curricula. He has designed and delivered webinars on emerging technologies such as embedded IoT, physical computing, mobile robotics concepts, and electronics prototyping methods. Currently, Dr. Don Wilcher is providing consultancy around virtual reality mechatronics, plant safety, and robotics course modules with industry standards alignment, writing technical and maker space project articles for EETech Media. He is actively involved in developing and delivering emerging engineering webinars for the Design News Continuing Education Center.

I want to thank my colleagues who have been close to me and supported me, and especially my wife Mattalene, my children, my friends, and my parents.

About the reviewer

Lensey C. King Jr. has worked in the electrical industry for three decades as a residential, commercial, and industrial electrician. He has received an Associate in Applied Science degree in automotive/automated manufacturing technology from Jefferson State Community College. Lensey is a licensed master electrician/electrical contractor for the states of AL, GA, and VA. He is working as a business development manager and partner of Multi-Skill Training Services in Alabama, where he is responsible for closing the "skills gap" in the manufacturing industries by offering customized industrial maintenance training. He is also the owner of EM&T, an electrical contracting company that focuses on maintenance service work and EV charging station repair.

Table of Contents

3

Lights, Sound, and Motion with M5Stack   71

Part 2: M5Stack Electronic Interfacing Circuit Projects

4

It's a SNAP! Snap Circuits and the M5Stack Core   103

5

Solderless Breadboarding with the M5Stack   131

6

M5Stack and Arduino   167

Part 3: M5Stack IoT Projects

7

Working with M5Stack and Bluetooth   199

8

Working with the M5Stack and Wi-Fi   227

Preface

Suppose you are interested in learning new approaches to developing portable electronic controllers to operate electromechanical loads such as a DC motor or introducing new prototyping techniques for interactive product development using littleBits or SNAP circuits as augmentation tools. In that case, the *M5Stack Electronic Blueprints* book is for you. This book provides coverage of the design, development, and prototyping workflow process of building interactive electronic controllers and IoT device applications. You will learn how to create a physical M5Stack Core controller device based on the ESP32 microcontroller using discrete electronic components, Arduino C/C++, and UIFlow Blockly code (no-code) programming languages.

The ESP32 subsystem is the heart of the M5Stack Core controller. Here, internal circuits, such as a 3W audio amplifier, LED bars, USB-C ports, push button switches, a **thin film transistor** (**TFT**) touch screen, and **general-purpose input-output** (**GPIO**) ports, are managed by the ESP32 subsystem. You will learn about the structure of the ESP32 subsystem architecture in this book. The ESP32 microcontroller has an internal Bluetooth chipset and Wi-Fi support circuitry. You will learn how to create IoT controllers and scanners using the M5Stack internal Wi-Fi and Bluetooth chipsets. Finally, **human computing interaction** (**HCI**)-based techniques for creating effective and aesthetically appealing user interfaces for product engagement will be presented in this book.

Who this book is for

The *M5Stack Electronic Blueprints* book topics will assist you in exploring electronic controllers, Bluetooth, Wi-Fi-based IoT device development, and Arduino interfacing techniques using an ESP32 microcontroller-based platform. The M5Stack Core allows ease in developing new concepts for prototyping automation controls, wearables, interactive electronic controllers, and investigating wireless product technologies.

Therefore, the audience for this book includes the following:

- STEM, engineering, and technical educators
- Practicing engineers
- Electronics, automation, production, and industrial maintenance technicians
- Makers
- STEM and career and technical learners
- Electronic circuit hobbyists
- Inventors and academic researchers

What this book covers

Chapter 1, Exploring the M5Stack Core, explains the inner workings of the M5Stack Core, ESP32 microcontroller subsystem architecture, and internal supporting hardware electronics.

Chapter 2, Exploring M5Stack Units, explores M5Stack units for control and detection. The units are small electronic circuits such as the IR remote, environment sensor, RGB LED, motion sensor, and angle sensor. The M5Stack units are vital components for developing wearable electronics. You will use the M5Stack UIFlow Blockly code and the Arduino IDE C/C++ code to investigate the function of the extendable electronic devices.

Chapter 3, Lights, Sound, and Motion with the M5Stack, explores small wearable devices, including an electronic flashlight, an emergency flasher, a tone generator, an interactive emoji, and a haptic controller. The M5Stack has several electronic devices such as a microphone, an **Inertial Measurement Unit (IMU)**, RGB LED bars, a vibration motor, and a speaker.

Chapter 4, It's a SNAP! Snap Circuits and the M5Stack Core, covers SNAP circuits, which are electronic parts mounted on colored plastic shapes. Each shape has snap elements that allow the construction of electronic circuits. You will learn how to enhance the SNAP circuits user experience with an M5Stack Core controller by providing interfacing controls to operate the snap-based electronic devices.

Chapter 5, Solderless Breadboarding with M5Stack, provides wiring instructions for using a solderless breadboard and discrete electronic components. Touchscreen controllers and electronic sensors will be created using the M5Stack Core controller and the UIFlow Blockly code programming language. You will build hardware devices such as an electronic flasher, littleBits LED flasher, and DC motor controller in this chapter.

Chapter 6, M5Stack and Arduino, explains how to use electronic interface circuit techniques for wiring the M5Stack Core 2 controller to an Arduino Uno. You will investigate using electronic interfacing circuits and portable touchscreen controls for operating and monitoring Arduino Uno electronic devices, such as a touch-control inverting switch, a touch-control digital counter, and a touchscreen LED dimmer controller.

Chapter 7, Working with the M5Stack and Bluetooth, looks at the embedded Bluetooth chipset that is part of the ESP32 microcontroller. You will build a wireless controller using the embedded Bluetooth chipset. You will explore a wireless transmitter and receiver to operate RGB LEDs, small DC motors, and sound/tone generators using the M5Stack Core Bluetooth chipset. You will learn how to work wireless devices using a smartphone or tablet and a mobile Bluetooth UART utility services app.

Chapter 8, Working with M5Stack and Wi-Fi, explains the Wi-Fi support circuitry integrated within the ESP32 microcontroller's system architecture. You will conduct Wi-Fi experiments to scan and detect wireless network nodes in this chapter. You will explore visual detection indicators and audible alarms. This chapter will introduce you to the use of the Arduino IDE and an API to program Wi-Fi-enabled detection devices.

To get the most out of this book

The basic skill set the reader is assumed to have includes the following knowledge characteristics:

- A desire to learn about the ESP32 microcontroller

- The ability to identify basic electrical and electronic component symbols such as resistors, capacitors, tactile push button switches, wires, LEDs, and transistors on circuit schematic diagrams

- Basic knowledge of the Arduino electronics prototyping platform

- Basic knowledge of the Arduino **Integrated Development Environment (IDE)**

- The ability to modify **Application Programming Interface (API)** code

- The ability to read electronic circuit schematic diagrams

- The ability to wire electronic circuits using a solderless breadboard

To gain new skills in interfacing circuits, wearables, and embedded controller development, you should have a background in reading electronic circuit schematic diagrams. You should also be comfortable with using solderless breadboards. Finally, you should be able to identify electronic components such as transistors, resistors, and electrical male and female pin header connectors.

Software/hardware covered in the book	Operating system requirements
UIFlow Blockly code	Windows, macOS, or Linux
Arduino IDE	
ECMAScript 11	

After installing the UIFlow Blockly code software, a reset of the development system is required to complete the installation process. Having knowledge and experience of using a digital multimeter will aid you in troubleshooting interfacing circuits and wiring concerns with the projects presented in this book.

If you are using the digital version of this book, we advise you to type the code yourself or access the code from the book's GitHub repository (a link is available in the next section). Doing so will help you avoid any potential errors related to the copying and pasting of code.

Download the example code files

You can download the example code files for this book from GitHub at `https://github.com/PacktPublishing/M5Stack-Electronic-Blueprints`. If there's an update to the code, it will be updated in the GitHub repository.

We also have other code bundles from our rich catalog of books and videos available at `https://github.com/PacktPublishing/`. Check them out!

Download the color images

We also provide a PDF file that has color images of the screenshots and diagrams used in this book. You can download it here: `https://packt.link/873X7`.

Conventions used

There are a number of text conventions used throughout this book.

`Code in text`: Indicates code words in text, database table names, folder names, filenames, file extensions, pathnames, dummy URLs, user input, and Twitter handles. Here is an example: "The name of the app is `M5GO.py`."

Bold: Indicates a new term, an important word, or words that you see onscreen. For instance, words in menus or dialog boxes appear in **bold**. Here is an example: "Select **System info** from the **Administration** panel."

> **Tips or important notes**
> Appear like this.

Get in touch

Feedback from our readers is always welcome.

General feedback: If you have questions about any aspect of this book, email us at `customercare@packtpub.com` and mention the book title in the subject of your message.

Errata: Although we have taken every care to ensure the accuracy of our content, mistakes do happen. If you have found a mistake in this book, we would be grateful if you would report this to us. Please visit `www.packtpub.com/support/errata` and fill in the form.

Piracy: If you come across any illegal copies of our works in any form on the internet, we would be grateful if you would provide us with the location address or website name. Please contact us at `copyright@packt.com` with a link to the material.

If you are interested in becoming an author: If there is a topic that you have expertise in and you are interested in either writing or contributing to a book, please visit `authors.packtpub.com`.

Share Your Thoughts

Once you've read *M5Stack Electronic Blueprints*, we'd love to hear your thoughts! Scan the QR code below to go straight to the Amazon review page for this book and share your feedback.

https://packt.link/r/1803230304

Your review is important to us and the tech community and will help us make sure we're delivering excellent quality content.

Download a free PDF copy of this book

Thanks for purchasing this book!

Do you like to read on the go but are unable to carry your print books everywhere?

Is your eBook purchase not compatible with the device of your choice?

Don't worry, now with every Packt book you get a DRM-free PDF version of that book at no cost.

Read anywhere, any place, on any device. Search, copy, and paste code from your favorite technical books directly into your application.

The perks don't stop there, you can get exclusive access to discounts, newsletters, and great free content in your inbox daily

Follow these simple steps to get the benefits:

1. Scan the QR code or visit the link below

https://packt.link/free-ebook/9781803230306

2. Submit your proof of purchase

3. That's it! We'll send your free PDF and other benefits to your email directly

Part 1: M5Stack Electronics Hardware Architecture

The learning objective for *Part 1* is to acclimate you to the M5Stack controllers' internal electronics hardware. To accomplish this goal, you will perform basic coding tasks using the UIFlow software. The tasks include setting up USB communications between the host development system and the M5Stack Core controller, UI basics, operating the internal RGB LED bars and speaker-tone generator, and exploring the electronic circuit unit modules.

This part has the following chapters:

- *Chapter 1, Exploring the M5Stack Core*
- *Chapter 2, Hands-On with M5Stack Units*
- *Chapter 3, Lights, Sound, and Motion with the M5Stack*

1

Exploring the M5Stack Core

You are the proud owner of a small and powerful microcontroller development platform called the **M5Stack Core**. The M5Stack Core uses the **Espressif ESP32** microcontroller to allow you to create a variety of programmable electronic gadgets and devices. You can build programmable gadgets such as touchscreen controllers, Wi-Fi-operated devices, wearables, robots, and portable electronic games using the M5Stack Core. In addition, with the M5Stack Core, you can add visually appealing graphics and sound to your electronic creations easily and effectively using the **Arduino Integrated Development Environment** and the **UiFlow** *Blockly* code programming environment.

By the end of this chapter, you will understand the ESP32 microcontroller and the internal supporting electronics hardware. You will understand the M5GO Core hardware architecture's inner workings. Finally, you will know about applying the internal support electronics hardware to create an interactive electronic widget device.

In this chapter, we're going to cover the following main topics:

- Welcome to the M5Stack Core
- Overview of the M5Stack Core hardware architecture
- UI design basics
- UiFlow overview
- Communicating with the M5Stack Core

Technical requirements

To engage with the chapter's learning content, you will need the **M5GO IoT Starter Kit** to explore the internal/external hardware electronics and software coding of your first basic application. The UiFlow code block software will be required to build and run the M5Stack Core application.

You will require the following:

- M5Burner

- The M5Stack Core

- The M5GO IoT Starter kit

- UiFlow Blockly code programming software

Here is the GitHub repository for supporting software resources: `https://github.com/PacktPublishing/M5Stack-Electronic-Blueprints/tree/main/Chapter01`.

Welcome to the M5Stack Core

In this section, you will learn the basic functions and features of the M5Stack Core. You will be able to perform the following tasks upon completing the lessons in this section:

- Power and reset the M5Stack Core

- Prepare the unit for programming

- Identify the three programmable user-interface buttons

- Identify the external ports for connecting electronic units

- Identify the USB port

- Identify the LED bars

- Select your stored applications

- Change the initial setup operation of the M5Stack Core

Upon completing this section, you will have a solid foundation of the external features and internal functions of the M55tack Core. Being able to complete these tasks will be important for programming and testing software applications that you create with your M5Stack Core. We will start the discussion with an overview of the M5Stack Core external features.

Powering and resetting the M5Stack Core

Turning on the M5Stack Core is quite easy to do. The M5Stack Core has a small red power button, as seen in *Figure 1.1*. The small red power button is located on the left side of the M5Stack Core. You turn on the M5Stack Core with a single press of the power button. You reset the M5Stack Core with a quick double press of the power button:

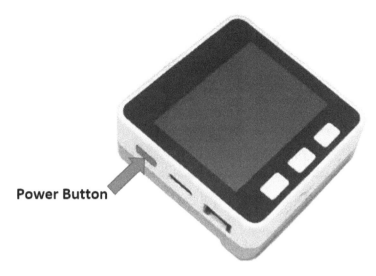

Figure 1.1 – The M5Stack Core power button location

Once the M5Stack Core is powered on, a UiFlow splash screen is displayed on the **liquid crystal display** (**LCD**), as seen in the following screenshot:

Figure 1.2 – UiFlow splash screen displayed after powering on the M5Stack Core

Initially, the M5Stack Core is programmed with 11 demonstration activities. The 11 demonstration activities are shown in the following diagram:

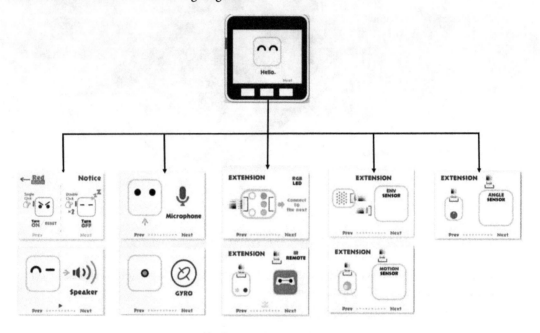

Figure 1.3 – UiFlow M5Stack Core demonstration activities

Next, let us learn about each demonstration activity through an accompanying explanation. We will explore the internal features of the M5Stack Core through a series of hands-on demonstrator activities:

- **Demo 1 on/off**: To turn off the M5Stack Core, a quick double press of the power button will accomplish this task, as seen here:

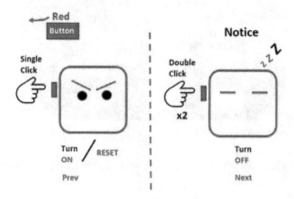

Figure 1.4 – The on/off operation of the M5Stack Core

- **Demo 2 speaker**: Press the center button on the M5Stack Core to hear a sound from the internal speaker:

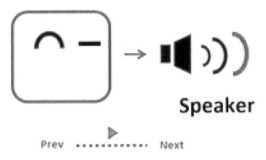

Figure 1.5 – Speaker demonstrator

- **Demo 3 microphone**: Speak into the pinhole microphone located on the side of the M5Stack Core. While speaking into the microphone, observe the line and sound wave images. Let's see how this looks in the following figure:

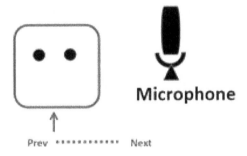

Figure 1.6 – Microphone demonstrator

- **Demo 4 gyro**: Tilt the M5Stack Core and observe the ball movement on the LCD screen. The movement is accomplished using a gyroscope and an accelerometer, as can be seen in *Figure 1.7*:

Figure 1.7 – Gyro demonstrator

- **Demo 5 RGB LEDs:** This demonstrator will glow and dim the sidebar RGB LEDs of the M5Stack Core. There is one LED bar on each side of the M5Stack Core:

Figure 1.8 – RGB bar demonstrator

- **Demo 6 ports explanation**: The different electrical ports and their specific functions will be explored in this demonstration:

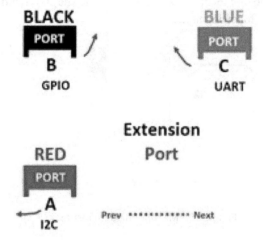

Figure 1.9 – Extension port demonstrator

Congratulations, you have completed exploring the internal features of the M5Stack Core. We will now learn about four sensor units from the M5GO IoT Starter Kit, which are shown in the following photo:

Figure 1.10 – The M5GO IoT Starter Kit

- **Demo 7 environment sensor**: The temperature and humidity levels are displayed on the LCD upon connecting the sensor to extension port A of the M5Stack Core:

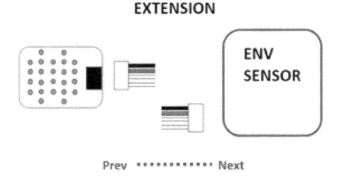

Figure 1.11 – Environment sensor demonstrator

- **Demo 8 passive infrared (PIR) sensor**: Attach the PIR sensor to extension port B of the M5Stack Core. You can observe a color change in the circle when placing your hand in front of the sensor. The color change response of the circle also occurs when you move your hand away from the PIR sensor. The following diagram illustrates the motion sensor's response to a hand moved away from the PIR sensor:

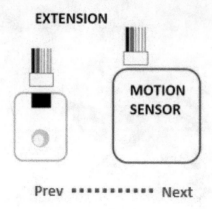

Figure 1.12 – PIR sensor demonstrator

- **Demo 9 RGB unit**: You explored the RGB LED bars in the fifth demonstrator activity. You will investigate the RGB LEDs packaged in this extension unit. This RGB LED unit will only light up when properly attached to extension port B. This diagram shows the wiring connector being attached to the RGB LED to connect to extension port B:

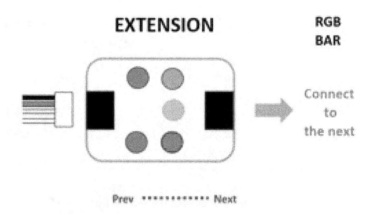

Figure 1.13 – RGB unit demonstrator

- **Demo 10 infrared (IR) Remote unit**: You can attach the IR Remote unit to extension port B using a wiring connector, as shown in *Figure 1.14*. Next, you take an ordinary IR handheld remote and point it toward the IR Remote unit. You can press any button on the IR handheld remote and observe the detection effect on the M5Stack Core LCD:

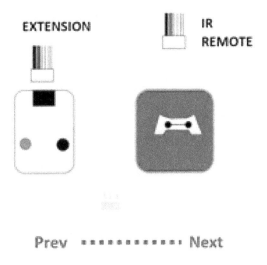

Figure 1.14 – IR Remote demonstrator

- **Demo 11 angle sensor**: You can test the knob on the angle sensor as shown in *Figure 1.15*, and observe the RGB LED bars getting brighter or dimmer:

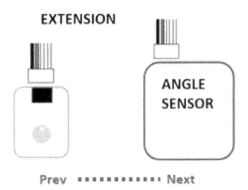

Figure 1.15 – Angle sensor demonstrator

Congratulations, you have learned about the M5Stack Core demonstrators. You have learned how to operate external sensors that are packaged with the M5GO IoT Starter Kit.

> **Note**
>
> To reinstall the firmware with the Demo app, use the M5Burner software. Select version 1.7.5 of the UIFLOW Core firmware. The name of the app is M5GO.py.

You have also learned how internal devices such as the RGB bar, microphone, and gyro operate using the appropriate M5Stack Core demonstrators. These mini hands-on demonstration activities allowed a glimpse into the M5Stack Core's hardware architecture.

In the next section, we will dive deeper into the M5Stack Core's hardware architecture by reviewing the electronic circuit schematic diagrams and the technical specifications of the ESP32-based programmable controller.

Overview of the M5Stack Core hardware architecture

In the previous section, you learned about the basic internal and external electronic features of the M5Stack Core. You investigated these basic features by using the M5Stack Demo app and specific IoT Starter Kit sensor units. In this section, you will learn about the embedded electronic subcircuits used to operate the M5Stack Core. The M5Stack Core's hardware architecture consists of the following four electronic subcircuits:

- Power management

- Audio amplifier

- ESP32 subsystem

- USB-UART and accessories

The M5Stack Core hardware architecture system block diagram is illustrated in *Figure 1.16*:

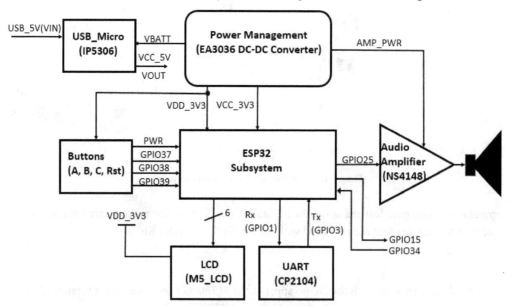

Figure 1.16 – M5Stack Core hardware architecture system block diagram

EA3036 DC-DC converter (power management)

The EA3036 **direct current to direct current** (**DC-DC**) converter is a power management **integrated circuit** (**IC**). The EA3036 IC can be powered by one **lithium-ion** (**Li-Ion**) battery. The input voltage range for the EA3036 is 2.7 **V** (**volt**) DC to 5.5 V DC. The IC can also be powered by a basic 5 V DC adapter or phone charger. The EA3036 integrates three synchronous buck converters into one convenient 20-pin **quad flat no-lead** (**QFN**) IC package. Let us learn more about the EA3036 DC-DC converter by reviewing the IC's physical pin packaging, shown in *Figure 1.17,* and the internal electronic circuits:

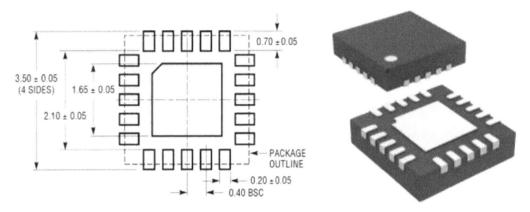

Figure 1.17 – EA3036 IC QFN package

The EA3036 DC-DC converter provides the 3.3 V (3V3) voltage source to operate the ESP32 microcontroller and support electronic circuit peripherals. Electronic circuit peripherals include the audio amplifier (N54148), A, B, C, and reset pushbutton switches, the LCD, the universal asynchronous receiver transmitter (CP2104) IC, and the USB-Micro (IP5306) circuit. The M5Stack Core Li-ion battery can power these subcircuits efficiently and effectively. The electronic circuit schematic and functional block diagrams for the EA3036 DC-DC converter are provided in the following diagram:

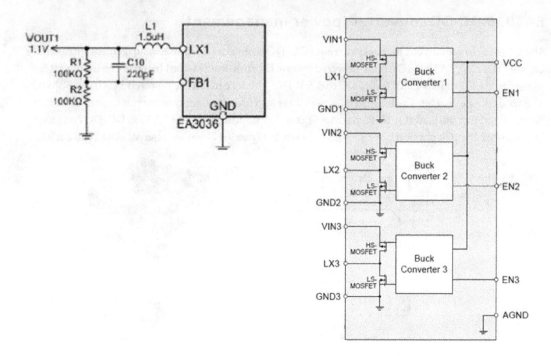

Figure 1.18 – The EA3036 DC-DC converter electronic circuit schematic and functional block diagrams

The EA3036 DC-DC converter can be used to source the appropriate voltage and current for the M5Stack Core's N54148 power amplifier.

Power amplifier

As illustrated in Demo 2, the M5Stack Core has an audio power amplifier. The NS4148 is a 3-**watt** (**W**) class D audio power amplifier. The unique feature of the NS4148 is that the device can power down, which reduces power consumption. This feature allows the M5Stack Core to maximize its 3.3V Li-ion battery capacity. The NS4148 audio power amplifier can be packaged in an 8-pin **micro-small outline package** (**MSOP**) or a small outline package. To minimize the printed circuit board space of the M5Stack Core, the MSOP IC package is used. *Figure 1.19* shows the MSOP of the NS4148 IC package:

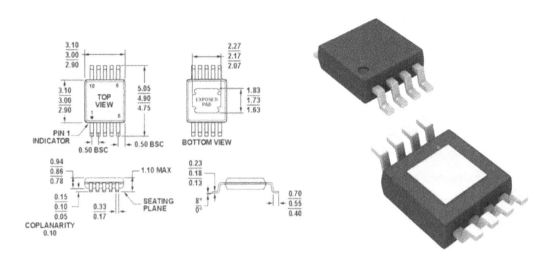

Figure 1.19 – The NS4148 MSOP device

The electronic circuit schematic for the NS4148 MSOP-based 3 W Class D audio power amplifier is shown in *Figure 1.20*. Although the NS4148 is a low **electromagnetic interference** (**EMI**) filter-less IC device, there is a pair of **inductor** (**L**) and **capacitor** (**C**) (**LC**) filters on the audio output pins. The LC filters consisting of **ferrite beads** (**FB1**), FB2, and capacitors C42 and C45 are to remove the amplified audio signal ripple noise or waveform distortion from being emitted to the M5Stack Core's speaker.

In addition, the bypass capacitor is connected across the **collector supply voltage** or **voltage common collector** (**VCC**) and **ground** (**GND**) pins to reduce high-frequency noise generated by the external clock circuit. The external clock circuit synchronizes the ESP32 microcontroller's internal timing operations, such as the program counter. A variety of sounds can be created using MicroPython, UiFLOW, or Arduino C/C++ coding languages. The NS4148 audio power amplifier will amplify these sounds generated by the ESP32 microcontroller, allowing the unique tones to be heard through the wired speaker. The NS4148 receives the ESP32 code-based tones from the GPIO25 pin.

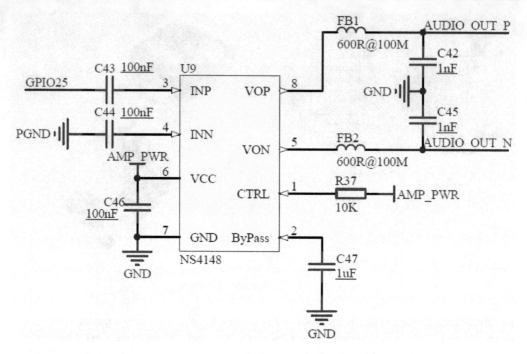

Figure 1.20 – The NS4148 3 W class D audio power amplifier circuit schematic diagram

ESP32 subsystem

The heart of the M5Stack Core is a 2.4 **gigahertz** (**GHz**) Wi-Fi and Bluetooth combination microcontroller. The ESP32 microcontroller is supported by 20 external components that enable features of the M5Stack to interact with the end user. The ESP32 provides appropriate control signals to the externally wired LCD and **universal asynchronous receiver-transmitter** (**UART**) devices. In addition, the A, B, C, and reset pushbutton switches are wired to the appropriate **general-purpose input/output** (**GPIO**) pins of the ESP32 microcontroller. The ESP32 microcontroller is packaged as a 48-pin QFN chip device. To ensure the proper timing of the ESP32's core processor functions as the program counter or shifting data through memory register movements, an external crystal clock circuit of 40 **megahertz** (**MHz**) is used.

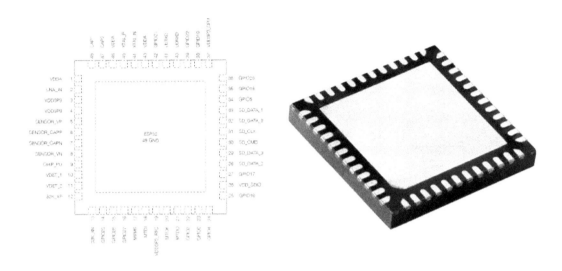

Figure 1.21 – ESP32 pinout and QFN48 package

The ESP32 microcontroller subsystem circuit schematic diagram is shown next:

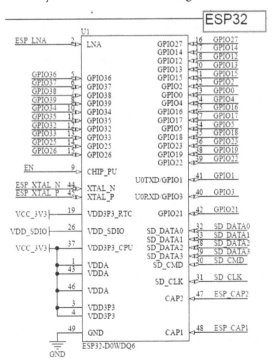

Figure 1.22 – ESP32 subsystem circuit schematic diagram

An external 40 MHz crystal clock circuit, as seen in *Figure 1.23*, is used to maintain the proper timing and data storage operations of the ESP32 microcontroller's core **central processing unit (CPU)**:

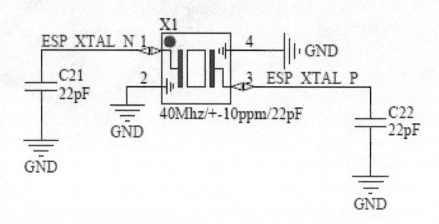

Figure 1.23 - A 40 MHz crystal clock circuit

USB-UART and accessories

The method used to communicate internally and externally with the M5Stack Core uses a **universal serial bus** (**USB**) or UART circuits. The external port A allows units with **inter-integrated circuit capabilities** (**I2C**) to communicate with the ESP32 microcontroller. A **serial clock** (**SCL**) and a **serial data address** (**SDA**) scheme allow such units to share data with the microcontroller. In addition, sending application programs created using UiFLOW Blockly code, MicroPython, or C/C++ programming languages is accomplished with a data **transmit/receive** (**TX/RX**) UART scheme through a USB communication port. An IP5306 circuit is used to accomplish these communication tasks. The two USBs depicted in the circuit schematic diagram in *Figure 1.24* represent the traditional data communication and extension A ports found on the M5Stack Core:

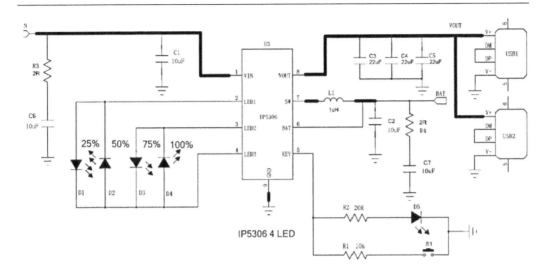

Figure 1.24 – The IP5306 USB-I2C communication circuit schematic diagram

Besides the internal circuits, the M5Stack Core uses a communication bus packaged on its bottom base to communicate with the microphone and the RGB LED bars. The communication bus allows the ESP32 microcontroller to receive and send audio and controls signals accordingly to the internal microphone and RGB LED bars. The allocated ESP32 GPIO pins are wired to the communication bus, thus allowing access to the audio and control signals from the externally connected devices. As shown in *Figure 1.25*, the GPIO15 pin operates the RGB LED bar. The ESP32's GPIO34 pin is wired to the microphone and amplifier circuit:

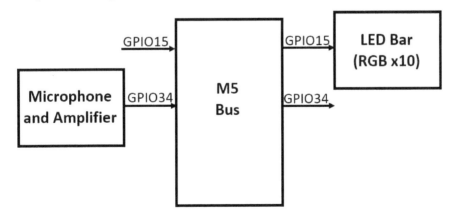

Figure 1.25 – The M5 communication bus and associated peripheral devices

The following photo shows some of the ESP32 microcontroller supporting ICs, the microphone, the RGB LED bar, and the M5 bus:

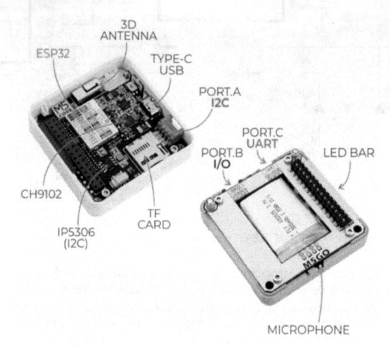

Figure 1.26 – Location of the key electronic components and M5 communication bus

You now understand the M5Stack Core's hardware architecture and the supporting electronic circuits. This knowledge will be important in the proceeding chapters of this book. The importance of this knowledge is in the development of interactive controls and wearable devices. In developing M5Stack Core projects, you will be immersed in the creative process of building practical and engaging products.

In the next section, you will learn how to design **user interfaces** (**UIs**) that allow interaction and engagement with the M5Stack Core.

UI design basics

In developing M5Stack Core applications, the UI aids in the operation of the device. The M5Stack Core's LCD easily lays out graphics that convey information from electronic sensors attached to the programmable controller. The M5Stack Core LCD can also provide the status as output devices like an electromechanical relay. In this section, you will learn about effective visual effect approaches for UIs. You will learn about UI design basics that make your applications appealing and easy to use. Before you start designing your UI, the dimensions of the M5Stack Core are needed.

The M5Stack Core's LCD dimensions are 320 x 230 pixels. Therefore, you have a total of 73,600 pixels available to create a variety of unique UIs:

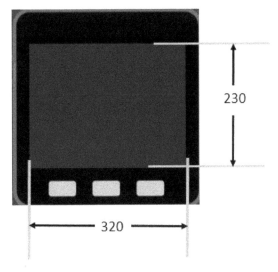

Figure 1.27 – M5Stack Core dimensions

The pixels are electronic dots arranged using a **Cartesian coordinate system**. When placing the images or controls, their locations can be managed using the Cartesian coordinate system. You can move the image quite easily using the UiFlow programming tool. The UiFlow programming tool has a design layout area that allows a preview of the images and controls placed on the LCD. The following example illustrates the x and y coordinates of 58 and 45 for the location of the square on the LCD:

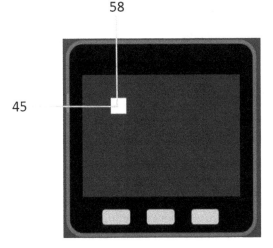

Figure 1.28 – Placement of an object on the UiFlow design layout area

The *x*-data point is 58 and the *y*-data point is 45. You can place the objects by dragging them to the specified location or entering them manually as follows:

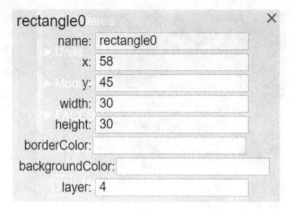

Figure 1.29 – Manual placement of a square on the UiFlow design layout area

You can obtain the properties box by clicking on the square. Once the properties box is visible on the UiFlow design layout area, *x* and *y* location data points may be entered as shown in the following figure:

Figure 1.30 – Entering x and y data points into the properties box

The square can be precisely located on the LCD by entering the value incrementally in the properties box. As you enter the value, the square will move accordingly based on the coordinate position. You can use this approach to lay out your UI design on the M5Stack Core's LCD. To make your UI effective, there are some basic design elements you need to consider.

Input controls

Input controls such as buttons, text fields, checkboxes, radio buttons, drop-down lists, list boxes, toggles, and date fields, allow the user to interact with your M5Stack Core device easily. You should size the input controls to be visible and for ease of use. Overlayed objects should not be placed on the input controls. Such information will make the input controls difficult to use.

Design considerations

When you lay out a UI design for an M5Stack Core device application, three factors should be considered: *development*, *visibility*, and *acceptance*. Let's look at them in detail:

- **Development factors**: These help improve the visual communications of the UI design by providing tool kits, component libraries, and rapid prototyping support. The M5Stack Core's main development factor is the availability of the UiFlow Blockly and Arduino code libraries. An example of a development factor is the Wi-Fi network UiFlow Blockly library shown in the following screenshot:

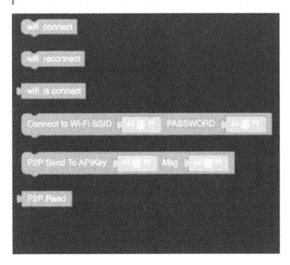

Figure 1.31 – Wi-Fi Blockly code library example of development factors

- **Visibility factors**: These allow multiple representations that aid the user to interact and engage with a UI design based on styles of learning and comprehension. Engagement, as it relates to the visual appeal of a UI design, provides education or is a productive tool and aligns with the learning aspect of the multiple representations concept. Examples of multiple representations include graphs, symbols, shapes, pictures, and sounds. The M5Stack Core includes a variety of shapes that will allow you to provide multiple representations for user interaction and engagement.

- **Acceptance factors**: These align with the documentation and training of the UI design-based product. For example, the M5Stack Core has been used in a variety of K-12 education classrooms based on the ease with which learners can create engaging and interactive devices. *Figure 1.32* illustrates an acceptance factor using multiple representations for learner engagement and device interaction:

Figure 1.32 – M5Stack Core UI shapes for multiple representation

A final important note in designing and developing M5Stack Core UIs is simplicity. Simplicity is the design consideration consisting of using the important UI elements for communicating features and functions of your M5Stack Core device. The UI you design should be simple, eliminating unnecessary input control elements. The labeling of the multiple representations you use should be clear in language and messaging. The following screenshot illustrates clarity, thus eliminating unnecessary input control elements:

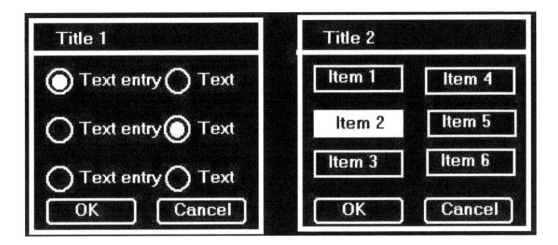

Figure 1.33 – (Left) complicated UI versus (right) simpler UI design

You now understand UI design basics that will aid in developing easy-to-use M5Stack Core applications. In the next section, we will present an overview of the UiFlow software.

UiFlow overview

In the previous section, you learned about some approaches to designing UIs to provide interaction for your M5Stack Core applications. The design basics provide suggested guidelines for developing uncluttered UI layouts. You can think of the design basics as the developmental theory to assist in creating UIs that have purposeful, engaging, and interactive functions for the M5Stack Core. In this section, you will learn about the UiFlow software basics. To start the learning journey, you will need to download UiFlow. You can use the following link to download the UiFlow software: `https://shop.m5stack.com/pages/download`.

Here's what the page looks like:

Figure 1.34 – M5Stack Core software

Download and install the **UIFlow-Desktop-IDE** (**integrated development environment**) software, as shown in *Figure 1.34*. Once the programming package has been installed, click on the icon to open the software, which looks like this:

Figure 1.35 – UiFlow icon

The UiFlow IDE will appear on your screen, as you can see in *Figure 1.36*. With the IDE open, various block categories, shapes, and an M5Stack Core preview will be visible. The UiFlow IDE will allow you to program the M5Stack Core using a variety of Blockly code blocks. You will construct your M5Stack Core application by stacking the Blockly code blocks and using shapes to create engaging product UI aesthetics and functions.

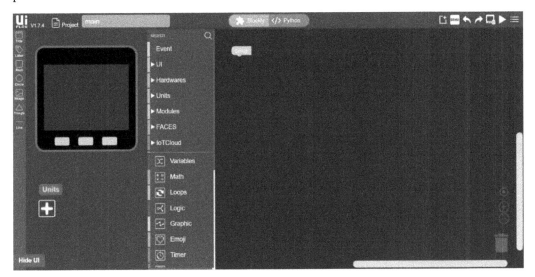

Figure 1.36 – UiFlow IDE

You will see that the UiFlow layout is very easy to understand and use. *Figure 1.37* explains UiFlow's IDE features. The M5Stack Core preview will allow you to design interactive devices to engage the end user with your product creation. To ensure your M5Stack Core device allows interaction and engagement with the intended product, you can consult the UI design basic material presented. A nice feature of the M5Stack Core preview is the ability to sketch out a UI design by using the shapes provided by the UiFlow software. You can proceed to design a UI that engages through simplicity and clarity.

Project Title
Enter your project name here. Whenever you download a program to your computer or M5Stack Core it will retain this name.

Blockl y</> Python
This button allows you to see the Python code that your blocks have generated and edit the code.

Menu Tab
From this bar you can access the forum, documentation, example undo and redo actions, upload files to the M5Stack Core, run your code on the device and alter the settings.

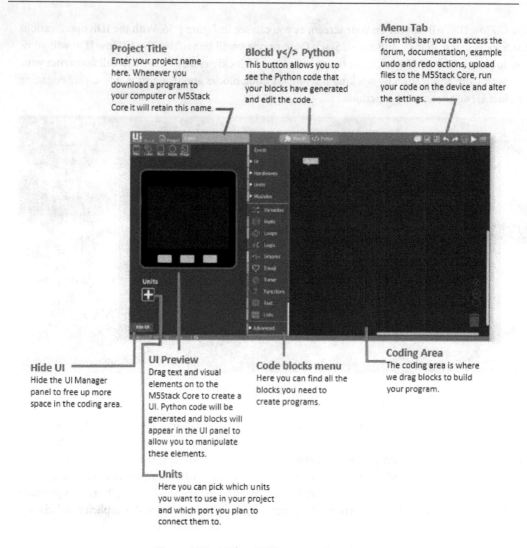

Hide UI
Hide the UI Manager panel to free up more space in the coding area.

UI Preview
Drag text and visual elements on to the M5Stack Core to create a UI. Python code will be generated and blocks will appear in the UI panel to allow you to manipulate these elements.

Code blocks menu
Here you can find all the blocks you need to create programs.

Coding Area
The coding area is where we drag blocks to build your program.

Units
Here you can pick which units you want to use in your project and which port you plan to connect them to.

Figure 1.37 – UiFlow IDE layout explanation

With the UI design sketched out using the M5Stack Core preview, you can proceed to add function to the aesthetics by selecting appropriate Blockly code blocks. The UiFlow IDE provides several programming code blocks to provide interaction and engagement with your intended product. The software approach used to code is a top-down method. You can think about using Blockly code blocks in the same way as creating a functional flowchart. Instead of using pseudocode to capture the logic of the M5Stack Core's function, the software algorithm development is created using Blockly code blocks. With the Blockly code blocks aligned with the UI design, you have now created an interactive device to engage your end user. There are several Blockly code categories to select within the coding development tool suite, as shown in the following screenshot:

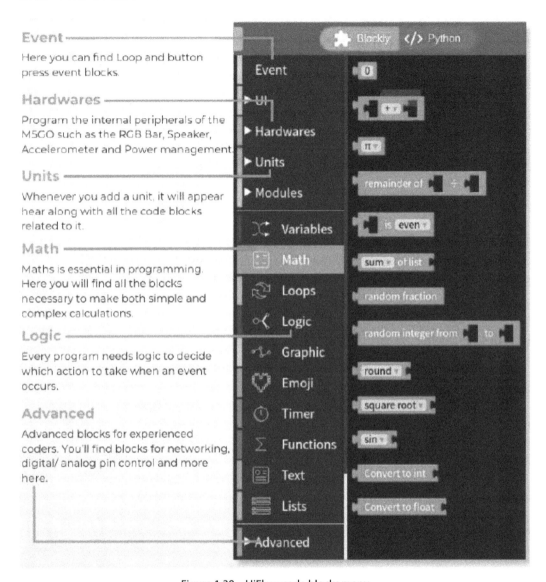

Event

Here you can find Loop and button press event blocks.

Hardwares

Program the internal peripherals of the M5GO such as the RGB Bar, Speaker, Accelerometer and Power management.

Units

Whenever you add a unit, it will appear hear along with all the code blocks related to it.

Math

Maths is essential in programming. Here you will find all the blocks necessary to make both simple and complex calculations.

Logic

Every program needs logic to decide which action to take when an event occurs.

Advanced

Advanced blocks for experienced coders. You'll find blocks for networking, digital/ analog pin control and more here.

Figure 1.38 – UiFlow code blocks menu

You can develop a Blockly code program as a sequence of interlocking instructions. Each code block functional instruction feeds and interlocks with the next program block. Therefore, a mental model can be created whereby creating M5Stack Core product functions align with a sequence of a program diagram. You can use the following diagram to develop simple or complex functions for your M5Stack Core device. You can practice using this mental model diagram when creating your M5Stack Core device's interactive function:

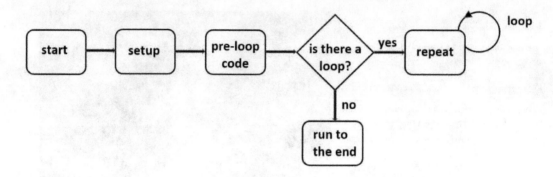

Figure 1.39 – UiFlow Blockly code program sequence diagram

Besides Blockly code block categories or menus, images, shapes, and the M5Stack Core preview, the menu tab plays an important role in the UiFlow software. The three basic Blockly code blocks that align with the program sequence diagram are **Setup**, **Loop**, and **Wait**, as shown in the following screenshot:

Setup

The setup block is essential for any program to run. It defines the first thing that will happen when the code is uploaded or the device is switched on. It will only run once.

Loop

The loop block will run any code placed inside it indefinitely . That means unless you turn off the device it will continue to run without stopping.

Wait

The wait block will delay your program for however many seconds you input. Sometimes this is necessary to see the result of some code that might have otherwise run so fast that you blinked and missed it.

Figure 1.40 – UiFlow Blockly code program sequence diagram

You will use these three Blockly code blocks quite often to allow the M5Stack Core device application's functions to be properly sequenced while maintaining engagement with the end user.

The role of the UiFlow menu tab is to provide you with a set of tools to run, test, and save your Blockly code. There is documentation on the code blocks and units for using them in your M5Stack Core device applications. Further, connecting to the M5Stack Core through USB can be achieved through the **Setting** menu. A description of the UiFlow Menu tab features is provided in *Figure 1.41*:

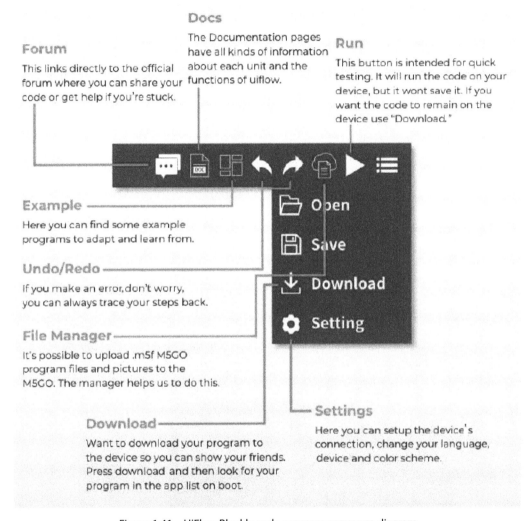

Figure 1.41 – UiFlow Blockly code program sequence diagram

The final section of this chapter will allow you to communicate with the M5Stack Core by establishing a proper communication setup with an ESP32-based device. You will use the UiFlow IDE software to communicate with the M5Stack Core device.

Communicating with the M5Stack Core

You are now familiar with the M5Stack Core's hardware architecture, UI design basics, and UiFlow IDE. In this section, you will learn how to set up the UiFlow IDE to communicate with the M5Stack Core. Being able to communicate with the M5Stack Core is important because the Blockly code applications you design and implement will now be tangible devices that people can use for controlling physical objects, such as LEDs, motors, and buzzers. You will be able to sense objects and environments using PIR and temperature sensors. Therefore, running software for these hardware components is dependent on communicating with the M5Stack Core device.

Requirements

To engage with this section's learning content, you will need the M5GO IoT Starter Kit to explore the internal/external hardware electronics and software coding of your first basic application. In addition, the UiFlow code block software will be required to build and run the M5Stack Core application.

You will require the following:

- The M5Stack Core controller
- The M5GO IoT Starter Kit
- UiFlow Blockly code software

Communication setup

The initial step to communicating with the M5Stack Core is to turn on the programmable device. Press the power **ON** button. Once the UiFlow splash screen is displayed, press the rightmost button on the M5Stack Core. The rightmost button is the M5Stack Core's **Setup** function.

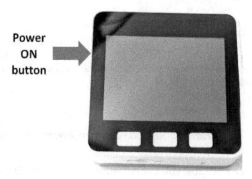

Figure 1.42 – Turning on the M5Stack Core

Then, perform the following steps:

1. Scroll down to **Switch mode**.

2. Scroll down to **USB Mode**. Use the middle button to select **USB Mode**.

These steps are shown in the following photo:

Figure 1.43 – Scroll to USB Mode

3. Scroll down to reboot the M5Stack Core. Use the middle button to select the **Reboot** mode. After selecting **Reboot**, the USB **API KEY** splash screen will be displayed on the M5Stack Core unit's LCD, as you can see in the following photo:

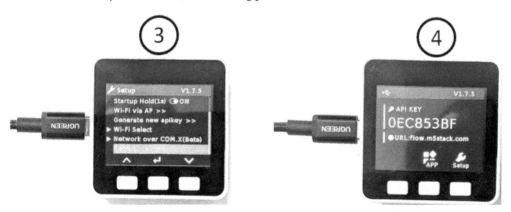

Figure 1.44 – Placing the M5Stack Core in USB mode (3 and 4)

You will then insert a USB-C cable into the M5Stack Core unit. Insert the other end of the USB-C cable into the development desktop personal workstation or laptop computer. Open the UiFlow Blockly code software. The **Setting** screen will appear on your personal desktop workstation or laptop screen. Select the M5Stack Core icon and the proper communication (**COM**) port, all of which you can see in the following screenshot:

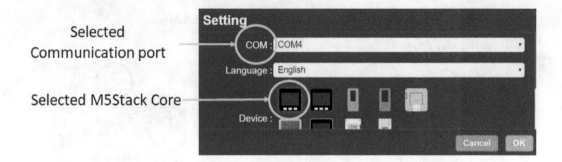

Figure 1.45 – Setting the communication port to select the M5Stack Core device

Once the M5Stack Core and communication port have been selected, click **OK**. The UiFlow IDE coding environment will be displayed on the screen as follows:

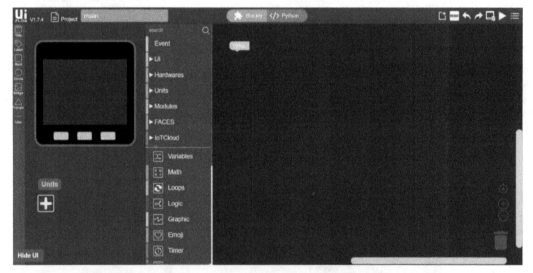

Figure 1.46 – UiFlow IDE

In the lower left-hand corner, you will see the **Connected** text displayed, as shown in the following screenshot:

Figure 1.47 – The M5Stack Core is connected to UiFlow Blockly code software

You will notice the **COM4** port is displayed with the **Connected** text, which confirms that the selection for the communication port was correct. Congratulations, your M5Stack Core is now communicating with the UiFlow IDE! In *Chapter 2*, you will learn how to attach and program small electronic modules such as angle and motion sensors, an RGB (red, green, and blue) LED, and an IR Remote unit to the M5Stack Core. The knowledge you have obtained in this chapter will be used to explore M5Stack Units in *Chapter 2*.

Summary

In this introductory chapter, you learned about the technical skills of understanding the M5Stack Core hardware architecture. Within the M5Stack Core's hardware architecture, you learned about the integrated circuits that provide the functional structure of this programmable device. You also covered how to operate the M5Stack Core demonstrators. Further, you learned about effective UI layout design approaches to make your M5Stack Core projects user-friendly. To implement the UI layout design approaches, you covered the UiFlow Blockly code features and functions that come as standard with the coding tool's IDE. Finally, you learned how to set up USB communications between the UiFlow Blockly code IDE and the M5Stack Core.

With this knowledge, you will be able to explore the M5Stack Core units in the next chapter. The M5Stack Core units are small electronic modules that work with the hardware architecture and Blockly code discussed in this chapter. You learned about the electronic circuits that aid in the hardware architecture of the M5Stack Core.

In the next chapter, you will use this electronic circuits knowledge, the design layout approaches, and the UiFlow IDE to program the units in Blockly code to create engaging and interactive devices.

2
Hands-On with M5Stack Units

You learned about the inner electronics of the **M5Stack Core** in *Chapter 1*. You also learned about UI design basics and received overview information on the UIFlow Blockly coding software. The new knowledge obtained in the previous chapter will allow you to learn about and use **M5Stack units** for creating interactive and engaging electronic devices.

In this chapter, you will explore M5Stack units for control and detection applications. The units are small electronic circuits such as an **infrared** (**IR**) remote, an angle sensor, an RGB LED, and a passive motion sensor. We will use the UIFlow Blockly code to investigate the function of the extendable electronic units.

By the end of this chapter, you will be able to perform the following technical tasks with M5Stack units:

- Draw a typical M5Stack unit connector electrical pinout
- Code an RGB LED unit UIFlow test application
- Code an IR remote unit UIFlow test application
- Code an angle sensor unit UIFlow test application
- Code a motion sensor unit UIFlow test application

In this chapter, we're going to cover the following main topics:

- Introducing M5Stack units
- Interacting with an RGB LED unit
- Interacting with an IR remote unit
- Interacting with an angle sensor unit
- Interacting with a motion sensor unit

Technical requirements

To engage with the chapter's learning content, you will need the **M5GO IoT starter kit** to explore the internal/external hardware electronics and software coding of your first basic application. The UIFlow code block software will be required to build and run the M5Stack Core application.

You will require the following:

- M5Burner software
- The M5Stack Core controller
- The M5GO IoT starter kit
- UIFlow code block software
- A USB C cable

Here is the GitHub repository for software resources: `https://github.com/PacktPublishing/M5Stack-Electronic-Blueprints/tree/main/Chapter02`

Introducing M5Stack units

The M5Stack unit is a small electronic input sensor or electrical output device that extends the interactive use of the M5Stack Core. There is a variety of units to select from to create a multitude of wearable and control applications. You can select from input units or output units. Input units take physical stimuli such as light or sound to create engaging detection devices. Output units such as RGB LEDs or buzzers provide visual and audio effects to your M5Stack Core creations. Examples of M5Stack units are shown in *Figure 2.1*:

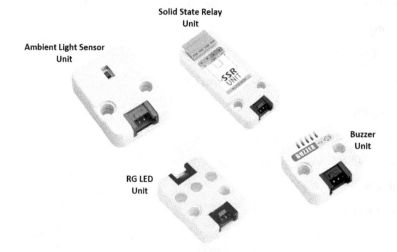

Figure 2.1 – Examples of M5Stack units

You attach the unit using a small jumper wire and mate it to the appropriate port. The M5Stack Core has three ports. The ports are labeled *A*, *B*, and *C*. The following figure shows the three ports of the M5Stack Core:

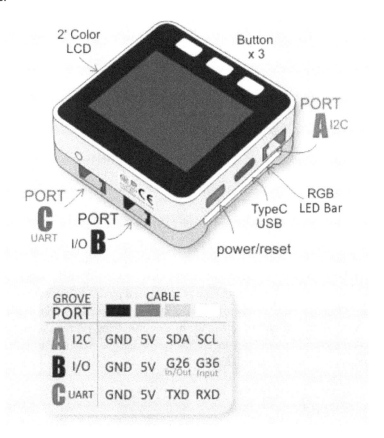

Figure 2.2 – M5Stack Core ports

You can use port A to handle basic digital input/output units along with inter-integrated circuits or I2C communications. Like port A, you can use port B with basic digital input/output units. You can use port C to communicate with other serial units, such as the unit CAM shown in *Figure 2.3*:

Figure 2.3 – Port C – Unit CAM application

As you can see, each individual port has four pins. All ports have a 5V and GND pin. The other two pins are specialized for each port. You select the unit based on the M5Stack Core application and the approach in which the unit works with the port's specialized pins. You can then create your own personal unit to perform the control or sensor task designed for the intended unique application. You can easily draw an electronic circuit schematic diagram showing your personalized unit wired to the M5Stack Core's electrical pinout. *Figure 2.4* shows an example electronic circuit schematic of personalized wiring for a push button switch:

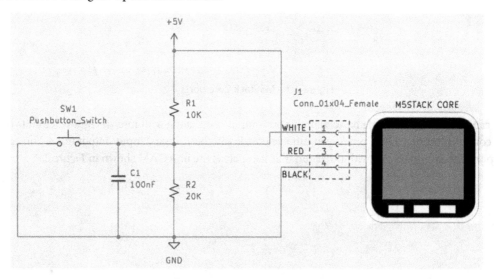

Figure 2.4 – Port B personalized wiring for a pushbutton switch

Now that you have learned about the various M5 Stack Core ports that allow different units to operate properly, you will build an RGB LED light using an RGB unit, a small jumper harness, and an M5Stack Core in the next section.

Interacting with an RGB LED unit

The RGB LED unit has three individual LEDs: red, green, and blue emitters. You can control each LED individually or simultaneously using code. You can also control the RGB LED unit's intensity. With the M5Stack Core, you can create visually appealing light effects using the RGB LED unit. *Figure 2.5* shows the RGB LED unit. To program the RGB LED, you will use the **UIFlow** Blockly code software discussed in *Chapter 1*.

Figure 2.5 – The RGB LED unit

The RGB LED unit consists of three individual pixels consisting of red, green, and blue colors. Each pixel is operated by an electronic control circuit. The colored pixels and electronic control circuit are packaged inside an RGB chip. The RGB chip is a **surface mount device (SMD)** of the 5050 variant component footprint. The **footprint** is the mechanical package of the electronic SMD device. You can see a picture and mechanical package dimensions of the RGB chip in *Figure 2.6*. The RGB chip's part number is **SK6812**:

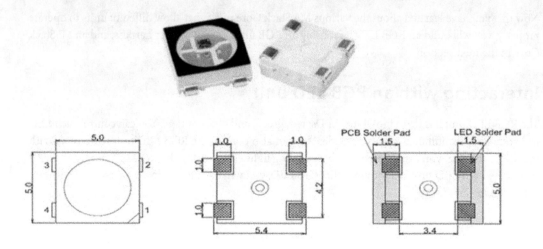

Dimensions in millimeters (mm)

Figure 2.6 – The SK6812 RGB chip and its mechanical package dimensions

To operate the RGB LED unit, you will attach the programmable device to the M5Stack Core using a four-wire jumper harness. You will use port B of the M5Stack Core to attach the RGB LED unit to the M5Stack Core. *Figure 2.7* shows the attachment of the four-wire jumper harness between the RGB LED unit and the M5Stack Core. The four-wire jumper harness has a color code scheme consisting of black, red, white, and yellow. You can verify that the jumper harness is correctly attached between the M5Stack Core and the RGB LED unit by the black wire located on the left side:

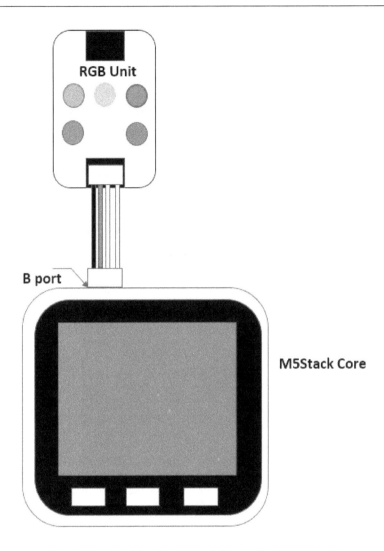

Figure 2.7 – Attaching the M5Stack Core to the RGB LED unit

You are now ready to program the RGB LED unit attached to the M5Stack Core using the UIFlow Blockly code software.

Programming the M5Stack Core to operate the RGB LED unit

You will use the UIFlow Blockly code software to program the M5Stack Core to operate the RGB LED unit. You will take a USB C cable and insert it into the M5Stack Core's equivalent port. The following figure illustrates the location and insertion of the USB C cable into the M5Stack Core:

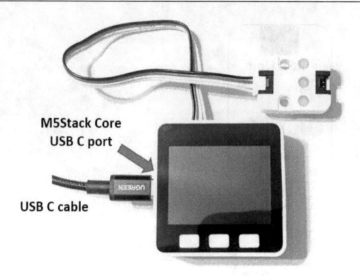

Figure 2.8 – Inserting a USB C cable into the M5Stack Core USB C port

You will take the other end of the USB C cable and insert it into your UIFlow Blockly code development machine. Your M5Stack Core should be turned on. You will press the middle button to configure your M5Stack Core to be in **USB mode**. *Figure 2.9* shows the M5Stack Core operating in USB mode.

> **Note**
> Refer to *Chapter 1* to configure the programmable controller for USB mode.

Figure 2.9 – The M5Stack Core operating in USB mode

You will then select from the appropriate Blockly code blocks bin to build the RGB LED unit application shown in *Figure 2.10*:

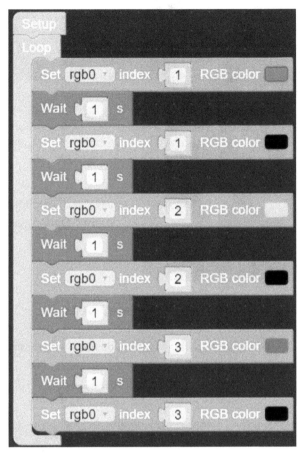

Figure 2.10 – RGB LED unit example UIFlow Blockly code

> **Note**
>
> The RGB LED unit needs to be selected so the programming blocks will be available for use in your application. At the lower left side of the UI layout, click the plus (+) sign with your mouse. Select the RGB LED unit to have access to the device's programming properties.

You will design the M5Stack Core's UI to display the text RGB example on the LCD. You will configure the UI text properties as shown in *Figure 2.11*:

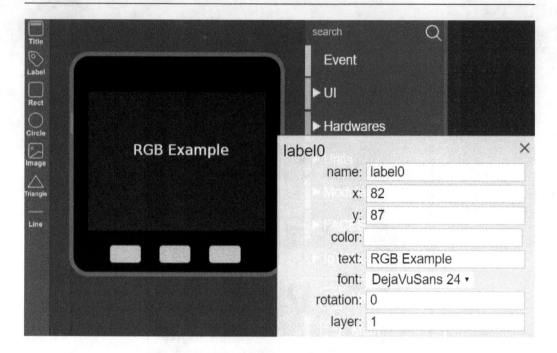

Figure 2.11 – The M5Stack Core's RGB LED UI display

You can run the RGB LED UI example by clicking the **Run** button with your mouse:

Run Button

Figure 2.12 – Executing the RGB LED UI example with the Run button

The Blockly code will run immediately on your M5Stack Core. *Figure 2.13* shows the green LED operating on the M5Stack Core. There is a 1-second delay between the M5Stack Core changing between the red, green, and blue LEDs. As a test, change the delay time to 500 **milliseconds (ms)** and observe the turn-on/off speeds of the LEDs.

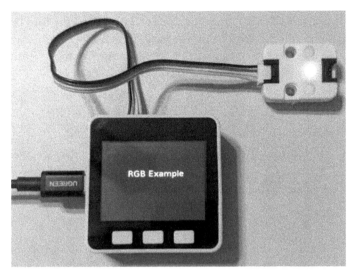

Figure 2.13 – The green LED turned on by the M5Stack Core

> **Interactive quiz 1**
> After changing the **Wait** block from 1 sec to 0.5 sec, did the RGB LEDs turn on slower or faster?

An interactive RGB LED unit

You can make the RGB LED unit interactive by pressing the *A* and *B* pushbuttons on the M5Stack Core. You will program the M5Stack Core *A* button to allow the red, green, and blue LEDs to turn on sequentially. You will program each color LED to turn on 500 ms. Conversely, you will program the *B* button to turn off the red, green, and blue LEDs. You will use the Blockly code program shown in *Figure 2.14*:

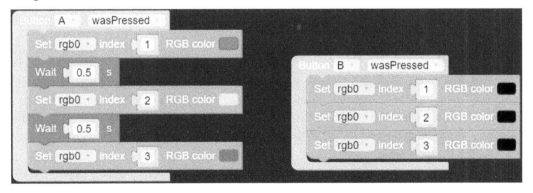

Figure 2.14 – Blockly code for an interactive RGB LED unit

You will run the interactive RGB LED unit Blockly code by pressing the **Run** button shown in *Figure 2.12*. When you press the *A* button on the M5Stack Core, the red, green, and blue LEDs will turn on sequentially. To turn the color LEDs off, you will press the *B* button. *Figure 2.15* illustrates the outcome of the interactive RGB LED unit Blockly code:

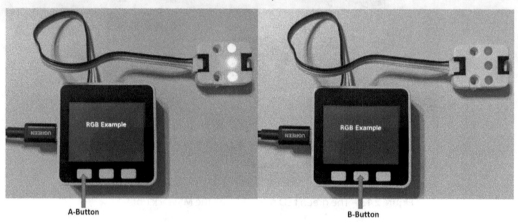

A-Button   B-Button

Figure 2.15 – An interactive RGB LED unit Blockly code running on the M5Stack Core

---

Interactive quiz 2

In reviewing the interactive RGB LED Blockly code shown in *Figure 2.14*, what is the index value for the green LED?

Congratulations! You have successfully programmed the M5Stack Core to perform some amazing control and UI interactive operations on the RGB LED unit. As demonstrated, laying out text for the M5Stack Core is quite easy and intuitive. In the next section, you will explore the IR remote unit using a handheld IR remote. You can use any available handheld IR remote to investigate the operation of the IR remote unit.

Interacting with an IR remote unit

The IR remote unit is an electronic photoelectric sensor capable of detecting infrared signals. The IR remote unit has an infrared **emitter** and **receiver** pair circuit packaged inside the electronic sensor unit. The IR remote unit is shown in *Figure 2.16*. The IR remote uses an IRM-3638 three-pin **integrated circuit (IC)**. As shown in *Figure 2.17*, the IRM-3638 IC is responsible for detecting and **decoding** or **demodulating** IR signal data.

Figure 2.16 – IR remote unit

Figure 2.17 shows a picture of the IRM-3638 IC:

Figure 2.17 – The IR Module (IRM-3638) IC

There is a photodiode used to transmit IR data from the M5Stack Core. The transmit ability of the IR remote unit provides the emitter operation of the photoelectric sensor device. Therefore, you can quite easily build a portable handheld IR remote tester using the M5Stack Core and this photoelectric sensor device. You will be exploring the IR remote unit in the next section with a basic demonstrator.

Programming the M5Stack Core to operate the IR remote unit

You will start the programming process of operating the IR remote unit using the M5Stack Core by electrically connecting the two devices with a four-wire jumper harness. *Figure 2.18* shows the attachment of the M5Stack Core to the IR remote unit:

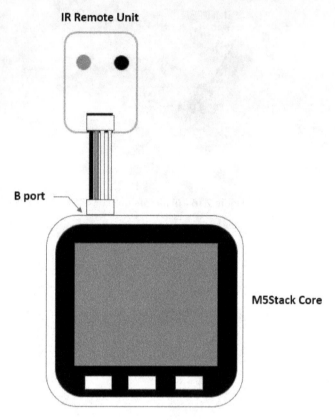

Figure 2.18 – Electrical attachment between the M5Stack Core and the IR remote unit

You will insert one end of the four-wire jumper harness into the M5Stack Core port B. You will take the other end of the four-wire jumper harness and insert it into the IR remote unit. The next step in this hands-on activity is to program the M5Stack Core to detect an IR signal from a handheld IR remote. The three important UIFlow Blockly code blocks enable the IR remote unit to detect infrared signals, and these blocks are shown in *Figure 2.19*:

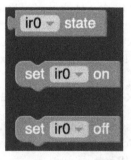

Figure 2.19 – Three important IR remote Blockly code blocks

The **ir0** state is the received IR signal. The numerical value returned signifying a detected IR signal is one (**1**). The **set on** Blockly code block transmits an IR signal. Finally, the **set off** Blockly code block stops sending an infrared signal. You will use these three key Blockly code blocks in various M5Stack Core IR remote applications. Next is a basic detection application using the discussed code blocks with the M5Stack IR remote and the M5Stack Core.

Basic IR handheld remote tester

You will build and test a basic IR handheld remote tester using the M5Stack Core and the IR remote unit. To build the tester, you will program and run the Blockly code software shown in *Figure 2.20* on the M5Stack Core:

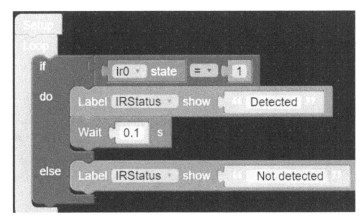

Figure 2.20 – IR signal detection Blockly code

The code will detect the state or presence of an IR signal. Upon detection, the M5Stack Core will display **Detected** on the **thin film technology** (**TFT**) LCD. With no IR signal detected, the TFT LCD will display **Not detected**. You will set the properties of the M5Stack Core's UI using the information shown in *Figure 2.21*. You will run the Blockly code on the M5Stack Core.

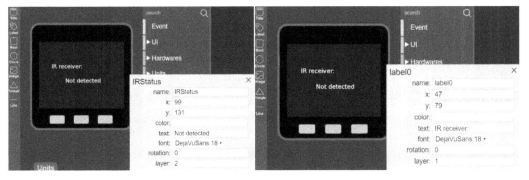

Figure 2.21 – The M5Stack UI properties setup

> **Note**
>
> Add the IR Remote unit to your UI layout design by clicking the plus (+) sign. Select the **IR Remote unit** icon. Now, you will have access to the IR remote unit's Blockly code blocks.

Place an IR handheld remote within proximity of the IR remote unit. Press any key on the IR handheld remote and observe the M5Stack Core's display. *Figure 2.22* illustrates the operation of the tester:

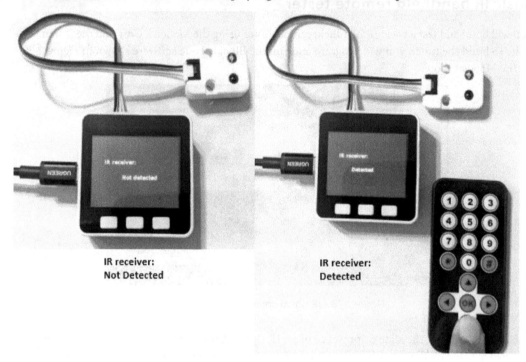

IR receiver:
Not Detected

IR receiver:
Detected

Figure 2.22 – IR remote unit detecting and handheld remote infrared signal

> **Interactive quiz 3**
>
> Using the Blockly code shown in *Figure 2.20*, change the **ir0** state value to zero and observe the operation. Did the detection response work in reverse?

Great job! You have built a programmable tester capable of testing various IR handheld remotes. Experiment with the UI's display message and program the *A* and *B* buttons to create unique testing and detection operations. Include sound to provide a unique tone when an IR signal is detected. In the next section, you will explore the operation of the angle sensor unit.

Interacting with an angle sensor unit

The angle sensor unit uses a basic 10 **Kiloohm (KΩ) potentiometer** for rotary adjustment of providing control signals. A potentiometer is a three-terminal electrical component that provides a range of resistance values. The primary function is to provide **voltage division** of an attached voltage supply source to an electrical circuit. *Figure 2.23* shows a typical potentiometer:

Figure 2.23 – A typical potentiometer

The angle sensor unit is shown in *Figure 2.24*. This potentiometer-based component attaches to the M5Stack Core's B-port:

Figure 2.24 – The angle sensor unit

To fully understand the angle sensor unit's potentiometer and voltage division, you can build an electronic circuit model using the original circuit schematic diagram, as shown in *Figure 2.25*:

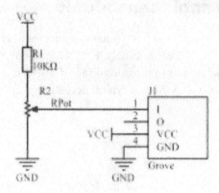

Figure 2.25 – The angle sensor unit circuit schematic diagram

The angle sensor unit circuit analysis

You can use Autodesk Tinkercad Circuits to create a functional breadboard version of the angle sensor unit. Autodesk Tinkercad Circuits is an online circuit simulation environment where various Arduino Uno and BBC micro:bit microcontroller platforms are used to build and test electronic circuit concepts. *Figure 2.26* shows a Tinkercad Circuits equivalent of the angle sensor unit circuit schematic diagram shown in *Figure 2.25*:

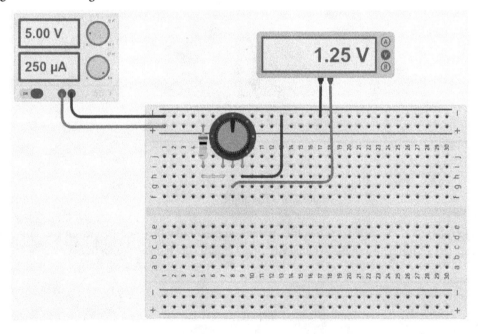

Figure 2.26 – Angle sensor unit Tinkercad circuit

You can use the following website address for tinkering with the Tinkercad circuit shown in *Figure 2.26*: `https://www.tinkercad.com/learn/circuits`.

As you can see in *Figure 2.26*, the potentiometer is set at 50% of the full 10 KΩ resistance value. The voltage measured from the potentiometer is 1.25V. This voltage is determined by multiplying 50% by the **direct current** (**DC**) voltage applied to the potentiometer. The voltage applied across the potentiometer is 2.5V. Therefore, 50% x 2.5V is equal to 1.25V. *Figure 2.27* shows the **voltage drop** across the potentiometer of 2.5V. The value of 50% can be written as ½, which is the fractional equivalent to the percentage number. Thus, the circuit is performing a division operation on the potentiometer's output voltage.

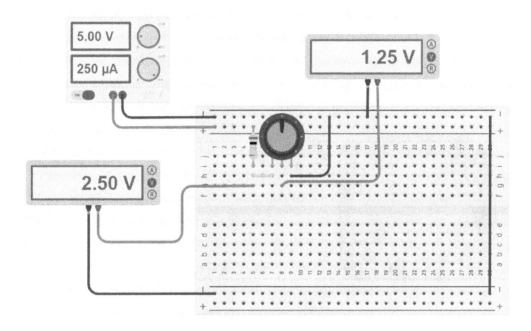

Figure 2.27 – Measuring the potentiometer output voltage drop

> **Note**
> Here is a simple equation to use when determining the potentiometer's output voltage: *Vout = percentage x potentiometer voltage drop*.

You can see the ease with which the potentiometer's output can be determined using a circuit simulation model and a basic equation. With such knowledge, you can build your own angle sensor unit using a 10 KΩ potentiometer and a fixed 10 KΩ resistor. *Figure 2.28* illustrates a Homebrew angle sensor unit you can build using the specified **passive** components.

> **Interactive quiz 4**
>
> What is the Vout with an adjusted potentiometer voltage percentage of 30%?

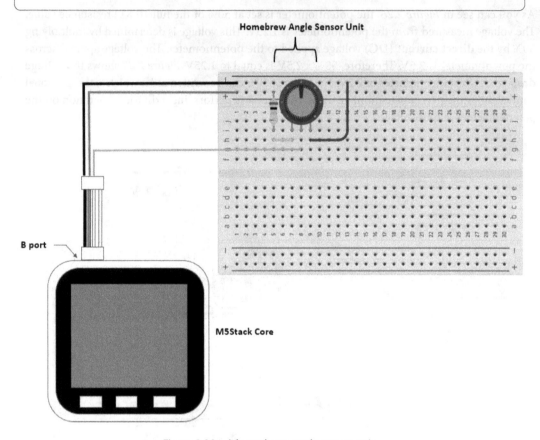

Figure 2.28 – A homebrew angle sensor unit

To obtain a greater understanding of this circuit analysis discussion, you will test the original angle sensor unit and the homebrew device in the next section. You will program the M5Stack Core to read the angle sensor and display an analog-to-digital converter value on the TFT LCD.

Programming the M5Stack Core to read the angle sensor unit

You will start the programming process of reading the angle sensor unit using the M5Stack Core by electrically connecting the two devices together with a four-wire jumper harness. *Figure 2.29* shows the attachment of the M5Stack Core to the angle sensor unit:

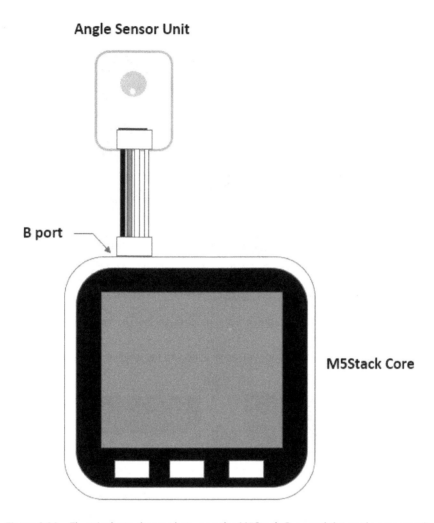

Figure 2.29 – Electrical attachment between the M5Stack Core and the angle sensor unit

You will insert one end of the four-wire jumper harness into the M5Stack Core B-port. You will take the other end of the four-wire jumper harness and insert it into the angle sensor unit. The next step in this hands-on activity is to program the M5Stack Core to detect and display angle rotation from the sensor. You can set up UIFlow to interact with the angle sensor unit by adding the potentiometer device to the programming environment. *Figure 2.30* shows the angle sensor included in the UI layout section of the UIFlow programming environment:

Angle Sensor unit

Figure 2.30 – Adding the angle sensor unit to the UIFlow UI layout section

By adding the unit to the UI layout section, you have access to the angle sensor unit Blockly coding block, as shown in the following figure:

Figure 2.31 – Angle sensor unit Blockly coding block

You can easily create an M5Stack Core display device that will display the equivalent angle value. The displayed angle numeric represents the unit's equivalent **analog-to-digital converter** (**ADC**) value. The maximum ADC value that can be displayed is 1,024. You will use the following Blockly code to read the angle sensor unit's equivalent ADC value and display it on the M5Stack Core's TFT LCD. The angle sensor unit's Blockly code is shown in *Figure 2.32*:

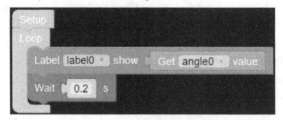

Figure 2.32 – Angle sensor unit ADC display Blockly code

You will set the properties of the M5Stack Core's UI using the information shown in *Figure 2.33*. You will run the Blockly code on the M5Stack Core:

Figure 2.33 – The M5Stack UI properties setup

You will see the minimum ADC value of **0.0** to the maximum value of **1024.0** displayed on the TFT LCD shown next. These values are displayed by adjusting the angle sensor unit's potentiometer. *Figure 2.34* illustrates the adjustment of the angle sensor and its equivalent ADC values:

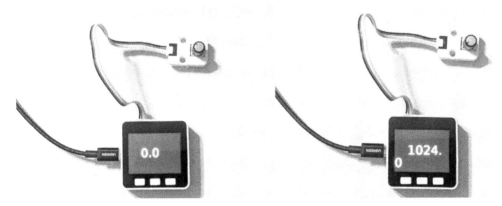

Figure 2.34 – ADC minimum and maximum values displayed on the TFT LCD

Note

The ADC value of **0.0** is equated to 0V and the ADC value of **1024.0** aligns with 5V.

Congratulations, you have created a UI appliance that displays ADC values using the angle sensor unit and the M5Stack Core. As a final lab project, you can create a small interactive numeric display using the main ADC value obtained from the angle sensor unit. You can use additional Blockly code blocks to create this small interactive numeric display. *Figure 2.35* provides the revised code for the small numeric interactive display:

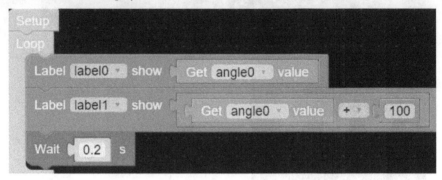

Figure 2.35 – Blockly code blocks for a small numeric interactive display

You will need to add a label (`label1`) to display the small numeric values produced by the angle sensor unit and Blockly code previously shown in *Figure 2.35*. You will configure `label1` using the properties shown in *Figure 2.36* to display the new numeric values:

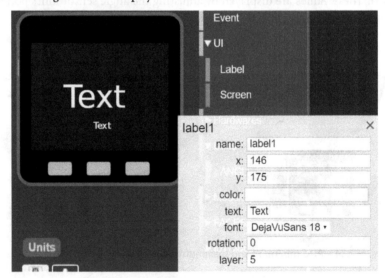

Figure 2.36 – label1 properties

With the code built and the label properties settings configured, you can run the code on M5Stack Core. You will see a small numeric display underneath the original text. The small numeric display is one-hundredth smaller than the original value. *Figure 2.37* shows the new M5Stack Core display:

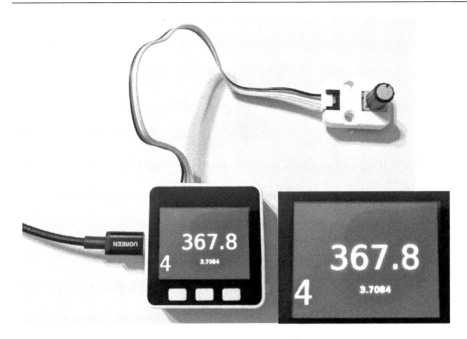

Figure 2.37 – M5Stack Core displaying an interactive small numeric display

> **Interactive quiz 5**
>
> Adjust the angle sensor unit's potentiometer to approximately 611 and observe the small numeric value displayed on the M5Stack Core's TFT LCD. Is the small numeric value displayed approximately one-hundredth of the original number?

Again, congratulations on creating this enhanced angle sensor unit demonstrator. In the final section of this chapter, you will investigate the inner workings of the motion unit. You will learn the basic operation of a **passive infrared** (**PIR**) sensing device through a hands-on activity. The hands-on activity will consist of attaching the motion unit with a four-wire jumper harness to the M5Stack Core and coding a basic human detection application. The cognitive knowledge and hands-on skills obtained in the previous sections will allow you to code and test a motion unit enabled by the M5Stack Core.

Interacting with a motion sensor unit

The motion sensor unit detects an object or a human's emitted heat energy using a passive **pyroelectric IR detector**. The device is **passive** based on no external power supply being needed to operate the detector. The pyroelectric crystal serves as the heat or IR detection element within the motion unit. When the pyroelectric crystal detects heat or IR, the surface of the crystal produces an electric charge. This electric charge is sent to an electronic switch that controls a visual signaling device, such as an LED. *Figure 2.38* shows the M5Stack motion sensor unit:

Figure 2.38 – An M5Stack motion sensor unit

> **Note**
>
> The change in a material's surface charge in response to temperature variations is known as the **pyroelectric effect**.

When the motion sensor unit detects an IR source, the device's output pin turns on. The output signal stays on for approximately 2 seconds. When the IR source is removed from the motion sensor unit, its output turns off. The motion sensor unit continuously monitors IR sources such as the human body or an object's heat energy. The main sensing element that is inside the motion sensor unit is a PIR AS312 component. The **PIR AS312 component** has a pyroelectric sensor that detects the human body or object's emitted IR energy. *Figure 2.39* illustrates an example PIR AS312 component:

Mechanical Package with Pinout

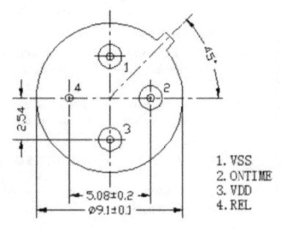

1. VSS
2. ONTIME
3. VDD
4. REL

Dimensions in millimeters (mm)

Figure 2.39 – A typical pyroelectric PIR AS312 component

The schematic diagram of the motion sensor unit is shown in the following figure:

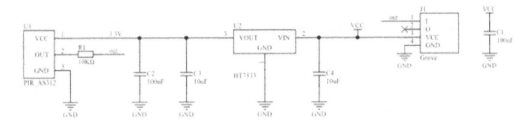

Figure 2.40 – The motion sensor unit's electronic circuit schematic diagram

The M5Stack Core assigned **general-purpose input-output** (**GPIO**) pinout is GPIO36. The pinout table shown next illustrates port B and the assigned GPIO36 pin:

M5Core(GROVE B)	GPIO36	GPIO26	5V	GND
PIR Unit	Sensor Pin		5V	GND

Figure 2.41 – The M5Stack Core assigned GPIO pin for the motion unit

In the next section, you will attach the motion sensor unit to the M5Stack Core using a four-wire jumper harness. You will then test the basic detection device with a simple PIR Blockly code application. This activity will check the electrical connection between the motion unit and the M5Stack Core using the PIR application.

Programming the M5Stack Core to detect a human body with the motion sensor unit

You will start the programming process of detecting the human body with the motion sensor unit and the M5Stack Core by electrically connecting the two devices together with a four-wire jumper harness. *Figure 2.42* shows the attachment of the M5Stack Core to the motion sensor unit. As shown in *Figure 2.42*, port B is used for the Motion Sensor unit to communicate with the M5Stack Core:

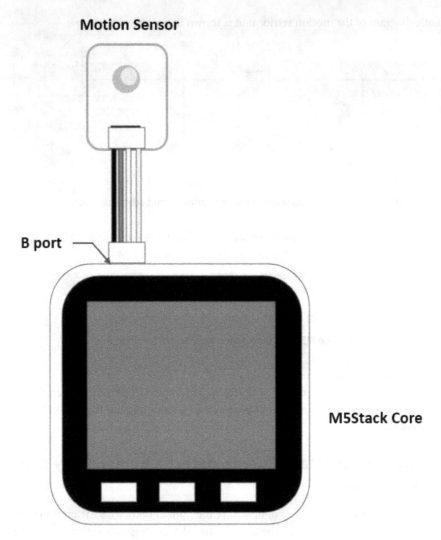

Motion Sensor

B port

M5Stack Core

Figure 2.42 – Electrical attachment between the M5Stack Core and the Motion Sensor unit

You will insert one end of the four-wire jumper harness into the M5Stack Core B-port. You will take the other end of the four-wire jumper harness and insert it into the Motion Sensor unit. The next step in this hands-on activity is to program the M5Stack Core to detect and display that a human body is present. You can set up UIFlow to interact with the motion sensor unit by adding the pyroelectric device to the programming environment. *Figure 2.43* shows the motion sensor unit included in the UI layout section of the UIFlow programming environment:

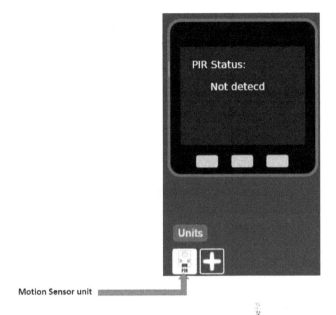

Figure 2.43 – Adding the motion sensor unit to the UIFlow UI layout section

By adding the unit to the UI layout section, you have access to the PIR unit Blockly coding block, as shown in *Figure 2.44*:

Figure 2.44 – PIR unit Blockly coding block

You can easily create an M5Stack Core display device that will display when a human body has been detected. The TFT LCD will initially display **PIRStatus: Not detected**. Upon detecting a human body, the TFT LCD will display **PIRStatus: Detected**. You will use the following Blockly code to read a human body's IR energy with the Motion Sensor unit. The Motion Sensor unit will display the detection message **Detected** when a human body is sensed. The Motion Sensor unit's Blockly code is shown in *Figure 2.45*:

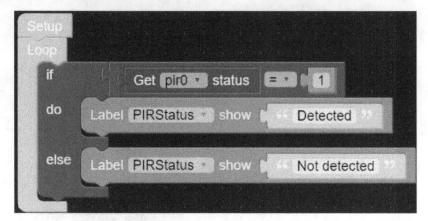

Figure 2.45 – Blockly code blocks for the motion sensor unit human body detector

You will set the properties of the M5Stack Core's UI using the information shown in *Figure 2.46* for label1:

Figure 2.46 – label1 properties

You will set the properties of the M5Stack Core's UI using the information shown in *Figure 2.47* for the PIRStatus label:

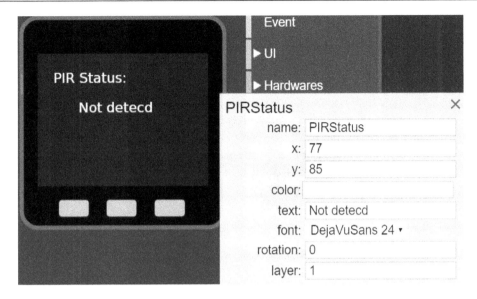

Figure 2.47 – PIRStatus label properties

You will run the Blockly code on the M5Stack Core. You will see that the PIR status initially displayed **Not detected** on the TFT LCD. As you approach the Motion Sensor unit, after approximately 2 seconds, the TFT LCD will display **Detected**. These status messages will continuously toggle based on the presence or absence of a human body within the Motion Sensor unit. *Figure 2.48* illustrates the toggling of the PIR status messages on the TFT LCD:

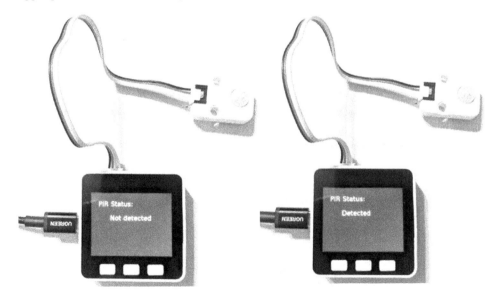

Figure 2.48 – Motion Sensor unit human body PIR Status detection

> **Interactive quiz 6**
>
> The sample code used a binary value of 1 assigned to the `Get pir0 status` code block. Change the binary value to 0 and observe the detection unit's response. Did the change produce the opposite response?

You have completed the motion sensor activities successfully and now understand the operation of a pyroelectric detection device. The interactive quiz presented will have allowed you to explore an inverted operation response of the motion sensor in detecting human bodies. In addition, the interactive quiz will have allowed you to check your knowledge using a hands-on assessment approach to learning about the basic operation of the motion sensor.

Summary

Congratulations, you have completed the hands-on activities and interactive quizzes in this chapter, in which you learned about the technology used in the M5Stack Core unit. You learned how to electrically connect the RGB LED, IR Remote, and angle and motion sensor units to the M5Stack Core using a four-wire jumper harness. Further, you learned about the key electrical or electronic sensing and visual component used with each unit. For the angle sensor unit, you learned how to wire a virtual device using the online Tinkercad Circuits website. You learned how the angle sensor unit's potentiometer produces a proportional output voltage based on the percentage of the supply voltage applied to the variable resistor. With this knowledge, you learned how to measure the potentiometer's percentage of output voltage using a Tinkercad Circuits multimeter.

Additionally, you learned how to add and set up each unit's UI properties within the UIFlow software. You learned how to test each M5Stack unit's basic operation using simple code blocks. With these UIFlow code block applications, you were able to gain an understanding of how to read and display each unit's electrical output properties on M5Stack Core's TFT LCD. You were able to test your coding knowledge by answering the interactive quizzes through hands-on investigation.

With this knowledge, you will be able to explore the M5Stack Core hardware in the next chapter. The M5Stack Core hardware is an electronic subcircuit that works with the ESP32 hardware architecture and the UIFlow Blockly code discussed in *Chapter 1*. You learned about the internal electronic subcircuits that aid in the ESP32 hardware architecture of the M5Stack Core.

You will use this electronic circuits knowledge, design layout approaches, and the UIFlow IDE to program the internal hardware in Blockly code to create engaging and interactive devices in the next chapter.

Interactive quiz answers

- *Interactive quiz 1*: The LED turned on faster

- *Interactive quiz 2*: The index value for the green LED is 2

- *Interactive quiz 3*: Yes, the detected message initially displays first

- *Interactive quiz 4*: The potentiometer output voltage (Vout) equals 0.75 or 750mV

- *Interactive quiz 5*: Yes, the message initially displays first

- *Interactive quiz 6*: Yes, the response will be inverted because of the change of the binary bit value

3
Lights, Sound, and Motion with M5Stack

In the previous chapter, we learned about the various units that extend the **M5Stack Core** operational functions. You learned about the various ports that allow units to attach to the M5Stack Core. You learned how personalized wiring for creating a push button switch can aid the construction of such an electrical device using discrete electronic components. *Chapter 2* also introduced you to the approach of using a four-wire **jumper harness** for attaching units to the M5Stack Core. Further, you learned how to program the M5Stack Core using the UIFlow Blockly code software to operate various units and design basics and received overview information on this software. The new knowledge obtained in *Chapter 2* will allow you to learn about and use the external and internal hardware of the M5Stack Core for creating interactive and engaging electronic devices. In this chapter, you will explore M5Stack Core external and internal hardware for lights, sound, and motion applications. The external and internal hardware are electronic circuits such as an external vibration motor, the internal speaker, and the side RGB LED bars. You will use the UIFlow Blockly code to investigate the function of these external and internal hardware electronic circuits.

By the end of this chapter, you will know how to perform the following technical tasks with M5Stack external and internal hardware:

- Draw a typical M5Stack *hardware block diagram*
- Create and code a *haptic* controller UIFlow-based device
- Create and code a tone generator
- Create and code an electronic flashlight
- Create and code an emergency flasher
- Create and code an interactive emoji

In this chapter, we are going to cover the following main topics:

- Introducing M5Stack Core hardware
- Using the M5Stack Core vibration motor unit
- Using the M5Stack Core speaker
- Coding an M5Stack Core RGB LED flasher
- Coding an M5Stack Core interactive emoji

Technical requirements

To engage with the chapter's learning content, you will need the **M5GO IoT starter kit** to explore the internal/external hardware electronics and software coding of your first basic application. A littleBits vibration motor, the proto module, and a LEGO mounting plate may be purchased to build a Homebrew haptic device. The UIFlow code block software will be required to build and run the M5Stack Core internal and external device applications.

You will require the following:

- The M5Stack Core vibration motor unit
- The M5Stack Core controller
- The M5GO IoT starter kit
- UIFlow code block software
- A littleBits vibration motor
- The proto module
- A LEGO brick adaptor
- A USB-C cable

Here is the GitHub repository for software resources: `https://github.com/PacktPublishing/M5Stack-Electronic-Blueprints/tree/main/Chapter03`.

Introducing M5Stack Core hardware

The M5Stack Core is supported by several internal hardware circuits. With these internal hardware circuits, you can create visual interactive devices that engage the human senses. You can access these internal hardware circuits using Blockly code. The internal hardware circuits you will engage with in this chapter are the amplifier speaker, the RGB LED bar, and the pushbutton switches. Besides these internal hardware circuits, you will build and code a mountable vibration motor. The mountable vibration motor provides a sneak preview of *Chapter 5*. In this hardware introduction, you will be introduced to electrical circuit interfacing by building the mountable vibration motor. To introduce the mountable vibration motor, here is a block diagram representation of the external hardware device:

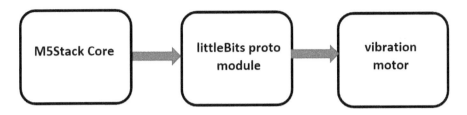

Figure 3.1 – Mountable vibration motor block diagram

As shown in *Figure 3.1*, each electronic component is represented by a block. The control signals flow from each block in the direction of left to right. The arrows represent the control signals flowing between the individual components. With this graphical approach, you can draw the hardware architecture of any wearable or electronic controller concept for the M5Stack Core. Here is another example illustrating the drawing of a block diagram:

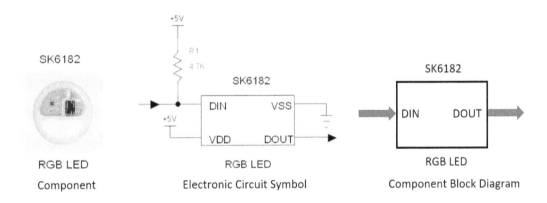

Figure 3.2 – RGB LED component block diagram

Figure 3.2 shows the RGB LED component, the **electronic circuit symbol**, and its block diagram. As you can see, the only elements captured in the block diagram are the input and output signal connections. Therefore, the purpose of the block diagram is to show the data or the signal flow between the interconnected hardware components. Thus, the block diagram shows the interconnected relationships between hardware components.

Interactive quiz 1

Using *Figure 3.1*, draw a block diagram showing an external pushbutton switch operating the vibration motor using the M5Stack Core.

Congratulations, you now have knowledge of drawing block diagrams for the M5Stack Core. You will use this knowledge for the projects in this chapter. You will now explore the vibration motor unit hardware using the M5Stack Core and the UIFlow Blockly code.

Using the M5Stack Core vibration motor unit

You can use the M5Stack Core to operate a vibration motor. The vibration motor unit has a weight that is offset and mounted to its rotating shaft. *Figure 3.3* shows an example of the M5Stack Core vibration motor, whose shaft has an attached metal-eccentric weight:

Figure 3.3 – M5Stack Core vibration motor unit

In general, a wide range of products use a vibration motor unit to provide physical stimuli to the user. This **physical sensory perception** or **haptic feedback** allows an individual to be totally immersed in the product through a human interaction experience. You can easily operate an M5Stack Core vibration motor using the UIFlow Blockly code software. In the next section, you will learn how to attach a vibration motor to the M5Stack Core. Further, you will learn how to program the M5Stack Core to operate the vibration motor.

Attaching the M5Stack Core to the vibration motor unit

The first step of programming the M5Stack Core to operate the vibration motor is the attachment of the **electromechanical component** to the programmable unit. To operate the vibration motor, you will attach the programmable device to the M5Stack Core using a four-wire jumper harness. You will use port B of the M5Stack Core to attach the vibration motor to the M5Stack Core. *Figure 3.4* shows the attachment of the four-wire jumper harness between the vibration motor and the M5Stack Core. The four-wire jumper harness has a color code scheme consisting of black, red, white, and yellow. You can verify that the jumper harness is correctly attached between the M5Stack Core and the vibration motor by the black wire located on the left side:

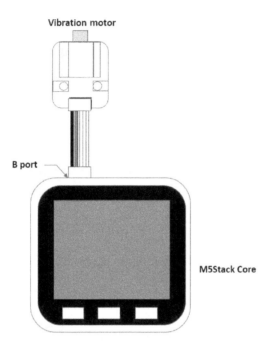

Figure 3.4 – Vibration motor unit attached to the M5Stack Core

You are now ready to program the M5Stack Core to operate the vibration motor using the UIFlow Blockly code software.

Programming the M5Stack Core to operate the vibration motor unit

You will use the UIFlow Blockly code software to program the M5Stack Core to operate the vibration motor. You will take a USB-C cable and insert it into the M5Stack Core's equivalent port. You will take the other end of the USB-C cable and insert it into your UIFlow Blockly code development machine. Your M5Stack Core should be turned on. You will press the middle button to configure your M5Stack Core to be in **USB mode**. *Figure 3.5* shows the M5Stack Core operating in USB mode.

> **Note**
>
> Refer to *Chapter 1*, in the *Communicating with the M5Stack Core* section. to configure the
> programmable controller for USB mode.

Figure 3.5 – The M5Stack Core operating in USB mode

You will select from the appropriate Blockly code blocks bin to build the vibration motor application
shown in *Figure 3.6*:

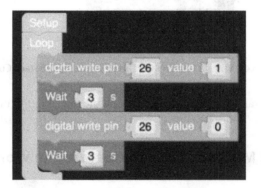

Figure 3.6 – Vibration motor UIFlow Blockly code

> **Interactive quiz 2**
>
> What effect would switching the binary values in the UIFlow Blockly code shown in *Figure 3.6*
> have on the vibration motor?

As the code is running on the M5Stack Core, the B port turns on and off every 3 seconds. The vibration motor initially turns on for 3 seconds. After the 3 seconds have expired, the vibration turns off and remains in that state for the same amount of time. When the motor is in the on state, the electromechanical component will vibrate, thus providing mechanical motion. Congratulations! You have successfully programmed the M5Stack Core to operate a vibration motor. With our success in operating the M5Stack Core vibration motor, let us take our knowledge of UI layout design and block diagrams to build a mountable vibration motor unit using littleBits electronic modules.

Building a mountable vibration motor unit

The concept behind this project is to create a mountable vibration unit using littleBits electronic modules and a LEGO brick adapter. *Figure 3.1* shows the block diagram for the prototype unit. The littleBits mountable vibration unit assembly diagram for attachment of the proto and vibration motor modules is shown in *Figure 3.7*. You will use the assembly diagram created from the block diagram to attach the littleBits electronic modules together:

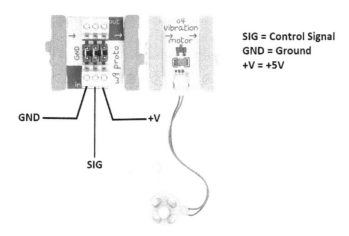

Figure 3.7 – Mountable vibration motor unit assembly diagram

To ensure the littleBits electronic modules are electrically stable, a LEGO brick adaptor is used. *Figure 3.8* illustrates the LEGO brick adaptor attached to the proto and vibration motor electronic modules:

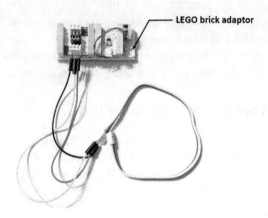

Figure 3.8 – LEGO brick adaptor attached to the mountable vibration motor unit

With the attached littleBits electronic modules assembled, you will wire the proto module to the four-wire jumper harness using the electrical wiring diagram shown in *Figure 3.9*. As illustrated in *Figure 3.9*, individual wires are used to make electrical extensions between the proto module and the four-wire jumper harness. These electrical extensions provide wire routing convenience for the mountable vibration motor. You can see that the vibration motor is magnetically attached to its snap leg. There are two small magnets on each side of the snap leg. The vibration motor will be safely secured to the snap leg illustrated in *Figure 3.9*.

> **Interactive quiz 3**
>
> With a digital voltmeter, what voltage would be sourced (measured) from the M5Stack Core's B port?

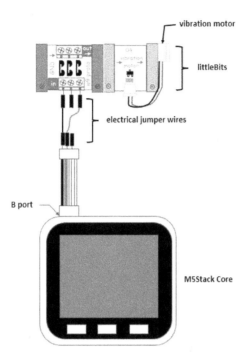

Figure 3.9 – Mountable vibration motor unit electrical wiring diagram

You will attach the complete unit to the back of the M5Stack Core using the LEGO brick adaptor. You will snap the LEGO brick adaptor to the back of the M5Stack Core by aligning the studs appropriately. *Figure 3.10* shows the mountable vibration motor unit attached to the M5Stack Core. Once the mountable vibration motor unit is attached, the final step of the project is to write the operational Blockly code.

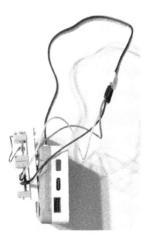

Figure 3.10 – The LEGO brick adaptor attached to the M5Stack Core

Programming the M5Stack Core to operate the mountable vibration motor unit

You will start the programming process of operating the mountable vibration motor unit using the M5Stack Core by electrically connecting the two devices with a four-wire jumper harness. You will take a four-wire jumper harness and attach the M5Stack Core and the mountable vibration unit together, as shown in *Figure 3.10*. Next, you will make the mountable vibration motor unit interactive by creating the UI layout as shown in *Figure 3.11*:

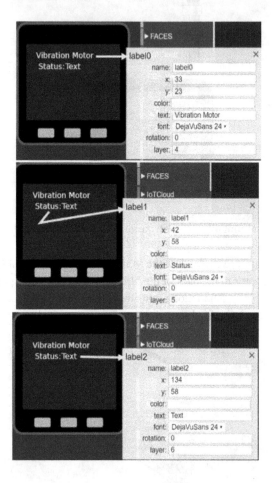

Figure 3.11 – UI layout for mountable vibration motor unit

You can make the mountable vibration unit interactive by pressing the *A* and *B* pushbuttons on the M5Stack Core. You will program the M5Stack Core *A* button to allow the mountable vibration motor unit to turn on. Conversely, you will program the *B* button to turn off the mountable vibration motor unit. You will use the Blockly code program shown in *Figure 3.12*:

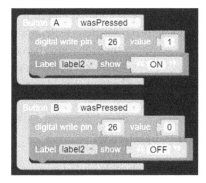

Figure 3.12 – Interactive mountable vibration motor unit Blockly code

Attach your M5Stack Core to the UIFlow development machine using a USB-C cable. Click on the **Run** button to download the interactive mountable vibration motor unit code to the M5Stack Core. Press the *A* button on the M5Stack Core to run the code. The littleBits vibration motor should be on. You stop the littleBits vibration motor by clicking the *B* button on the M5Stack Core. *Figure 3.13* illustrates the mountable vibration motor in operational mode:

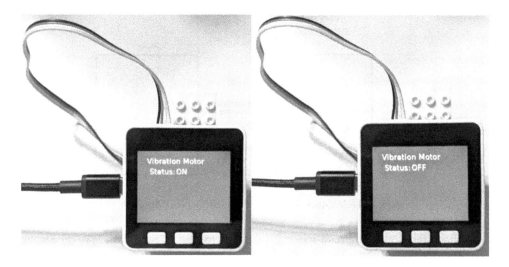

Figure 3.13 – An operational mountable vibration motor unit

Interactive quiz 4

What instructional UIFlow Blockly code block may be used to toggle the output state of the mountable vibration motor unit?

Congratulations, you have created a mountable vibration motor unit! Turn on the unit and hold it in your hand. The vibratory action produced is synonymous with haptic devices. **Haptics technology** allows the immersion of the user with an object through touch perception. Hand video controllers use a vibration motor to create an immersion experience between the gamer and the physical stimulus of the virtual world they are engaged in. You will explore the M5Stack Core's internal speaker in the next section.

Using the M5Stack Core speaker

The M5Stack Core speaker allows the programmable unit to be transformed into a portable creative sound generator. The M5Stack Core speaker is driven by a powerful 3W **class D amplifier**. The benefit of using this type of amplifier is its **efficiency**. Theoretical class D amplifiers have an operational efficiency of 100%. Typically, a class D amplifier has an efficiency of 90%. This high-efficiency rating is achieved by switching the amplifier's output transistors instead of driving them linearly. A **pulse width modulation (PWM)** signal effect is created, thus reducing the output transistors' thermal dissipation. Further, this output switching produces a high voltage and current amplification signal. With a reduction in **thermal dissipation**, the efficiency of the output transistors driving a speaker is highly effective. *Figure 3.14* shows a block diagram of the M5Stack Core speaker-amplifier circuit:

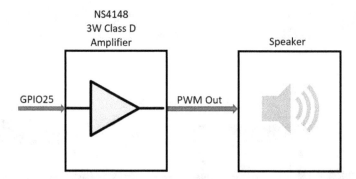

Figure 3.14 – M5Stack Core speaker-amplifier block diagram

Figure 3.15 shows the electronic circuit schematic diagram for the M5Stack Core speaker-amplifier block diagram:

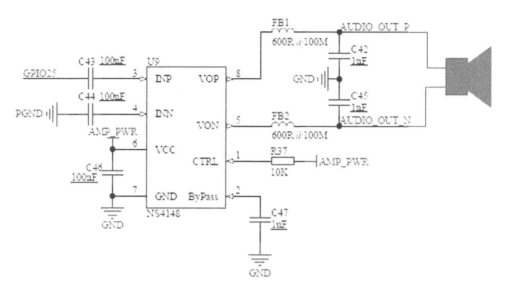

Figure 3.15 – M5Stack Core speaker-amplifier circuit schematic diagram

In reviewing the electronic circuit schematic diagram, you will notice an input coupling capacitor C43 with a capacitance value of 100nF. This C43 capacitor has an impedance value. This impedance value establishes the input **alternating current** (**AC**) resistance necessary to allow audio generator electrical sound to flow into the amplifier without signal reduction or attenuation. This impedance value is called **capacitive reactance** (**Xc**). You can calculate this equivalent AC resistance or impedance value with the following equation:

$$x_c = \frac{1}{2\Pi f c}$$

Figure 3.16 – Capacitive reactance equation

The variables are defined as such: f is the input signal frequency, c is the input capacitor value, π (pi), and Xc is the capacitive reactance in ohms. To illustrate how to use this equation, here is an example problem.

Determine Xc for the input capacitor shown in *Figure 3.15*. The input signal's applied frequency is 1.8 **kilohertz** (**kHz**) or 1,800 Hz.

The solution steps are shown in *Figure 3.17*:

$$x_c = \frac{1}{2\Pi f c}$$

$$x_c = \frac{1}{2\Pi_{1800100 \times 10} - 9}$$

$$x_c = 885\Omega$$

Figure 3.17 – Solution for the example problem

Therefore, the capacitive reactance for the M5Stack Core amplifier with a generated sound signal of 1,800 Hz is 885 Ω.

> **Interactive quiz 5**
>
> Using the solution steps shown in *Figure 3.17*, determine Xc for an M5Stack Core-generated sound signal of 100 Hz.

You now understand the type of amplifier that operates the M5Stack Core speaker. You also know the basic function of the class D amplifier, its block diagram configuration, and the equivalent **electronic circuit schematic diagram**. You have mathematical knowledge of calculating the capacitive reactance of an audio amplifier. In the next section, you will learn how to program the M5Stack Core to produce a basic sound. Further, you will learn how to make an interactive sound generator with a graphically displayed speaker icon.

Programming the M5Stack Core to produce a sound

You will start the programming process of producing a sound using the M5Stack Core by programming a basic warning alarm. You will operate the basic warning alarm by pressing the *A* button. The warning alarm will sound five times and stop. *Figure 3.18* shows the warning alarm block diagram:

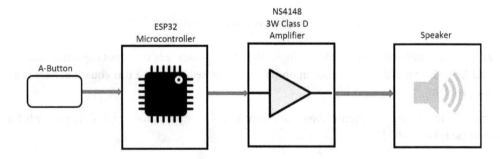

Figure 3.18 – The warning alarm block diagram

You will use the sound Blockly code blocks to program the M5Stack Core to produce an audible warning tone. *Figure 3.19* shows the location of the sound code blocks within the UIFlow programming palette. The selected sound blocks will be added with the *A* button and the `Wait` code blocks to create a two-tone alarm.

Figure 3.19 – The sound code block palette

The UIFlow Blockly code that aligns with the warning alarm block diagram is shown in *Figure 3.20*. You will use these code blocks to program the M5Stack Core. Click the **Run** button with your mouse to run the Blockly code on the M5Stack Core.

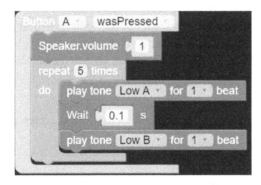

Figure 3.20 – The warning alarm Blockly code

Once you have built the code in the UIFlow environment, a simple speaker message can be added to display on the M5Stack Core TFT LCD using a **label**. *Figure 3.21* shows the UI label properties of the speaker message. As you can see, the label is somewhat placed in the middle of the M5Stack Core TFT LCD:

Figure 3.21 – The warning alarm M5Stack Core UI

With the Blockly code built and the UI designed, you can execute the Blockly code on the M5Stack Core by clicking the **Run** button with your mouse. Press the *A* button to hear a two-tone sound playing from the M5Stack Core speaker. Congratulations, you have successfully built a handheld warning alarm. You will now enhance the UI by adding a speaker icon to the M5Stack Core TFT LCD.

> **Interactive quiz 6**
>
> In reviewing the Blockly code shown in *Figure 3.20*, how can the interval time between tones be increased to 1 second?

Adding a speaker icon to the TFT LCD

You can make the speaker application TFT LCD appealing by adding an icon. The icon enhances the UI by visually identifying the operation of the programmed M5Stack Core. To add a speaker icon or picture to the M5Stack Core, drag the **Image** icon from the side panel and place it on the M5Stack Core's UI manager. *Figure 3.22* illustrates this action:

Figure 3.22 – Placing the Image icon onto the M5Stack Core UI manager

You will then configure the properties of the speaker label and the image. *Figure 3.23* illustrates configuring the label and image properties.

> **Note**
> The .jpg or .png file extensions are acceptable formats for displaying images on the M5Stack Core TFT LCD.

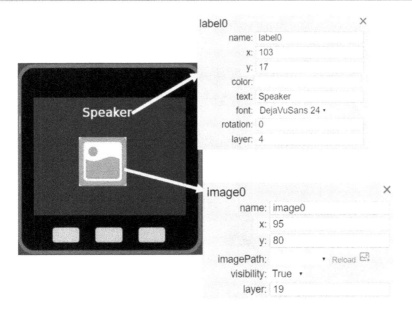

Figure 3.23 – Configuring the label and image properties

The speaker icon is uploaded to the image from a stored location on your desktop personal computer or laptop computer's hard drive. The `imagePath` field is the location of the stored speaker icon on your desktop personal computer or laptop computer's hard drive. You will click the **Reload** image to upload the speaker icon to the image. Once completed, the speaker icon should be displayed on the M5Stack Core's UI manager, as shown in *Figure 3.24*:

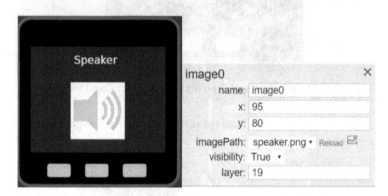

Figure 3.24 – The speaker icon displayed on the M5Stack Core's UI manager

Here is the speaker icon displayed on an actual M5Stack Core. The Blockly code shown in *Figure 3.20* may be used with this speaker icon-enhanced display:

Figure 3.25 – An M5Stack Core displaying a speaker icon

Congratulations, you have created an enhanced speaker application with a visually appealing TFT LCD. In this section, you have learned the basics of creating audible tones using the internal NS4148 3W **Class D amplifier** and speaker. In the next section, you will build a portable emergency flasher using the M5Stack Core's internal RGB LED hardware.

Coding an M5Stack Core RGB LED flasher

The M5Stack Core has internal RGB LEDs that can be commanded by the ESP32 microcontroller. The RGB LEDs are of the SK6812 programmable family variant. The SK6812 is a smart control circuit that operates an LED. The smart control circuit and light-emitting circuit are packaged inside one programmable LED source. You can refer to *Figure 3.2* for viewing the physical package, mechanical dimensions, and electronic symbol of the SK6812 RGB component. Another added feature of the SK6812 LED is the ability to emit a natural white light. Therefore, the SK6812 is an RGBW light-emitting component.

> **Note**
>
> Obtaining the natural white color requires a technique in color mixing of the RGB emitters. The following website provides information on obtaining white and various colors: `https://txwes.libguides.com/c.php?g=978475&p=7075536`.

The M5Stack Core uses two SK6812 LED bars. The SK6812 LED bar is located on each side of the M5Stack Core. *Figure 3.26* shows the location of one SK6812 LED bar. There are five SK6812 LEDs wired together to make one LED bar. Therefore, the M5Stack Core uses 10 SK6812 LEDs to allow various light patterns to be created by you.

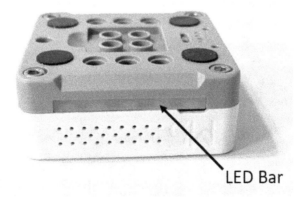

Figure 3.26 – An SK6812 LED bar

To create an RGB LED flasher, let us create a block diagram for the light-emitting switch device. The RGB LED flasher **block diagram** will consist of three components, a pushbutton switch, the ESP32 microcontroller, and the SK6812. *Figure 3.27* illustrates the RGB LED flasher block diagram. You will use the block diagram to code the RGB flasher in the UIFlow Blockly code coding environment. The RGB LED flasher code will be programmed onto the M5Stack Core.

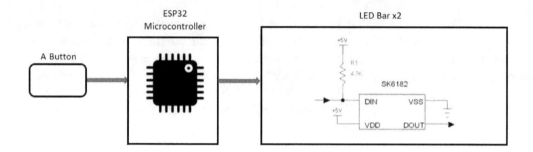

Figure 3.27 – M5Stack Core RGB LED flasher block diagram

Programming the M5Stack Core to operate the RGB LED as a flashlight

You will start the programming process of operating the IR remote unit using the M5Stack Core by obtaining the Blockly code blocks from the RGB code block palette. *Figure 3.28* shows the various RGB Blockly code blocks available for the M5Stack Core. As you can see, there are seven unique Blockly code block instructions available for the M5Stack Core to operate the LED bars:

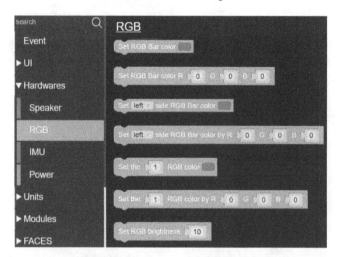

Figure 3.28 – RGB Blockly code block palette

Before building the RGB LED flasher, you will code a basic LED flashlight. To code this application on the M5Stack Core, you will use the Blockly code blocks shown in *Figure 3.29*:

Figure 3.29 – Basic LED flashlight code blocks

To adjust the red, green, and blue LEDs for a specific color, you can adjust the value using the slider control, as shown in *Figure 3.30*:

Figure 3.30 – Adjusting RGB colors using the slider control

You will adjust the green and blue values to **255**. The final code block is shown in *Figure 3.31*:

Figure 3.31 – The final Blockly code blocks for the M5Stack Core LED flashlight

> **Note**
> Besides using the slider control, you can enter the values directly into the textbox.

When you click the **Run** button in the top-right panel, the LED bars will turn on. You will notice that the color display is white. Therefore, your LED flashlight will emit white light. The value for obtaining white for the RGB parameters is 255. *Figure 3.32* shows the white light being emitted from an M5Stack Core's LED bars:

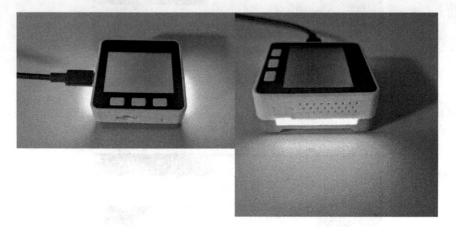

Figure 3.32 – M5Stack Core LED bars emitting white light

You will now code the *B* button to turn off the LED bars. The Blockly code blocks that will accomplish the function of your M5Stack Core LED flashlight are shown next:

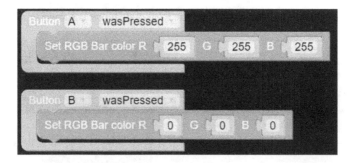

Figure 3.33 – Completed LED flashlight Blockly code blocks

Run the code and press the *A* button, and the LED bars will turn on. Now, press the *B* button, and the LED bars will turn off. Congratulations, you have created a flashlight using an M5Stack Core. The final step in this coding project is to make an RGB LED flasher. The next section will provide the Blockly code blocks to build the RGB LED flasher.

Interactive quiz 7

Using the LED flashlight Blockly code blocks shown in *Figure 3.33*, draw a block diagram for them.

Programming the M5Stack Core to operate the RGB LED as an LED flasher

In this final project, you will code the M5Stack Core to flash the RGB LED bars to run until the repeat value is reached. The **A** button starts the RGB LED bars' flashing cycle. The *repeat loop* code block will run a set of code block instructions. When the *repeat loop* code block has completed the instructions for operating the RGB LED bars, the flashing effect will stop. To track the *repeat loop* code block instructions, a counter is included. The counter keeps track of every on-and-off flash cycle. The timing between flashing the RGB LED on and off is 1 second. Therefore, it takes 2 seconds to complete one flashing cycle. The complete LED flasher Blockly code blocks are shown in *Figure 3.34*.

> Note
> Counters are used in various applications requiring events to be sequenced properly.

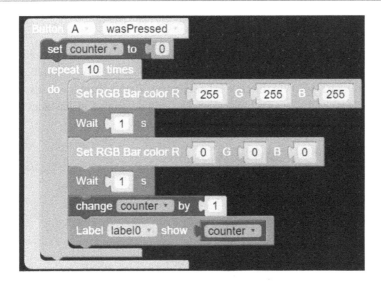

Figure 3.34 – RGB LED flasher Blockly code

To build the counter, you will need to create a variable. You can create the `counter` variable by accessing its code block palette. *Figure 3.35* shows the palette of Blockly code blocks available when the `counter` variable is created:

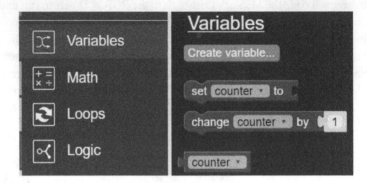

Figure 3.35 – The counter Blockly code palette

To display the counter on the M5Stack Core, a label placed on the TFT LCD is required. You configure the label using the techniques discussed previously in this chapter. The label will initially display **Text** on the M5Stack Core's TFT LCD. As the code is being executed, the counter will display its numeric value on the TFT LCD as the flasher switching cycle is sequenced. *Figure 3.36* illustrates the UI manager layout and properties for displaying the RGB LED flasher counter:

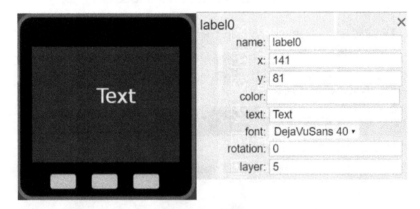

Figure 3.36 – The counter's UI manager layout and properties

With the code built, execute the application by clicking the **Run** button in the top-right panel. You will see the word **Text** follow by the LED bars flashing in 1-second intervals. The counter increments will be visible on the M5Stack Core's TFT LCD. *Figure 3.37* illustrates the RGB LED flasher's operation. You will notice, in reviewing the UIFlow Blockly code shown in *Figure 3.34*, that the counter increments every 2 seconds.

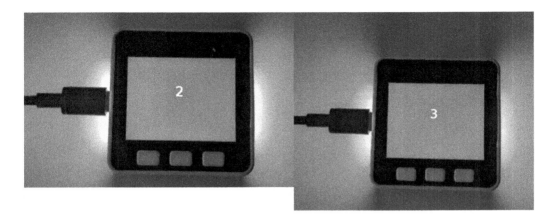

Figure 3.37 – The final RGB LED flasher operation

Interactive quiz 8

By changing the counter index value from 1 to 2 in the RGB LED flasher counter Blockly code shown in *Figure 3.34*, what starting count value will be displayed on the M5Stack Core TFT LCD?

Congratulations on completing the RGB LED flasher project! You have obtained knowledge of operating the internal M5Stack Core LED bar hardware using the UIFlow Blockly coding environment. You also have knowledge of using the `counter` and `repeat` Blockly code blocks to establish predefined flash rates for the LED bar hardware. In the final topic of this chapter, you will explore emojis and Blockly coding approaches to adding **human-computer interaction** to your M5Stack Core devices.

Coding an M5Stack Core interactive emoji

An emoji is a small icon that conveys the feelings of an individual in electronic mail or a document. With today's computer graphics capabilities, emojis have evolved into **interactive** and **animated** forms of **digital expression**. You will learn how to create a basic emoji using the M5Stack Core. The UIFlow Blockly coding environment has a palette of `Emoji` blocks. With these coding blocks, you can create animated and interactive emojis. *Figure 3.38* shows the `Emoji` code block palette:

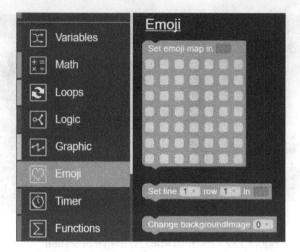

Figure 3.38 – Emoji code block palette

The primary Emoji code block is the Set emoji map in instruction. The approach to using the Emoji code block is to select the pixel or square of interest by clicking it with your mouse. By selecting appropriate pixels, a **two-dimensional** (**2D**) image can be created. The Change backgroundimage code block allows a predefined background to be present with your emoji. There are six predefined backgrounds that you can select for your emoji. To begin the interactive emoji project, you will create a smiley face emoji. *Figure 3.39* illustrates the pixels to select for the smiley face emoji:

Figure 3.39 – Smiley face emoji

With the appropriate code blocks placed on the UIFlow coding area, you will click the **Run** button to display the emoji on your M5Stack Core's TFT LCD. You will see a smiley emoji, as shown in *Figure 3.40*. The background is the default setting of 0.

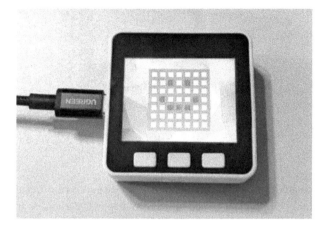

Figure 3.40 – M5Stack Core smiley emoji

You will create a sad emoji using the same approach that you used to make a smiley face. Here are the code blocks to create the sad emoji:

Figure 3.41 – M5Stack Core sad emoji code blocks

With the appropriate code blocks placed in the UIFlow coding area, you will click the **Run** button to display the emoji on your M5Stack Core's TFT LCD. You will see a sad emoji, as shown in *Figure 3.42*. The background is the default setting of 0.

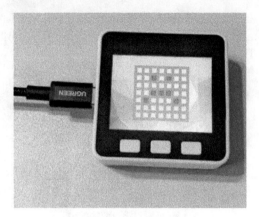

Figure 3.42 – M5Stack Core sad emoji

With two emoji blocks created, you will combine them to create a free-running animated emoji. You will use a `Wait` code block to provide a 1-second delay in displaying a smiley and sad emoji on the M5Stack Core TFT LCD. The code to accomplish this animation control function is shown in *Figure 3.43*:

Figure 3.43 – Emoji animated control code blocks

With the appropriate code blocks placed in the UIFlow coding area, you will click the **Run** button to display the animated emoji on your M5Stack Core's TFT LCD. You will see the smiley and sad emojis toggling, and the default background setting of 0 will be displayed. To add an interactive control feature, the *A*, *B*, and *C* buttons will be used. The features for each of the buttons are as follows: the *A* button will toggle the smiley and sad emojis, the *B* button will select the smiley emoji, and the *C* button will display the sad emoji. *Figure 3.44* shows the final interactive emoji Blockly code blocks:

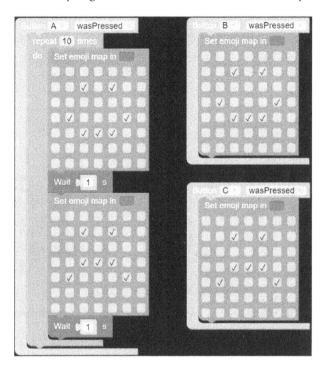

Figure 3.44 – The final interactive emoji Blockly code blocks

With the appropriate code blocks placed in the UIFlow coding area, you will click the **Run** button to execute the interactive emoji application on your M5Stack Core. You will see the animated emoji running on the M5Stack Core's TFT LCD when the *A* button is pressed. The smiley emoji will be displayed by pressing the *B* button. Pressing the *C* button will show the sad emoji. The background is the default setting of 0. Congratulations on completing the final project in the chapter!

Summary

Congratulations, you have completed the hands-on activities and interactive quizzes in this chapter! In the chapter, you learned about the circuit technologies used in the M5Stack Core. You learned how to draw block diagrams representing the M5Stack Core hardware architecture. You learned how to make a haptic controller using the M5Stack Core, a littleBits vibration motor module, and the hardware proto module. With the sound code blocks, you learned how to make a tone generator for a simulated warning alarm. You also explored the M5Stack Core's LED bars. In the hands-on exploration, you were able to make an electronic flashlight and an RGB LED flasher. These projects were accomplished with an RGB code blocks palette. In addition, you used your coding knowledge to answer the interactive quizzes through hands-on investigation. Lastly, you explored emojis in this chapter. You created an interactive emoji using the emoji code blocks within the UIFlow coding area.

With this knowledge, you will be able to explore the M5Stack Core as a programmable controller to operate external electronic circuits in the next chapter. The M5Stack Core input and output hardware ports will allow external electronic circuits to be operated with the M5Stack Core. You will use your knowledge of the M5Stack Core's subcircuits and hardware architecture to assist in wiring and controlling external electronic circuits.

Interactive quiz answers

1. Interactive quiz 1:

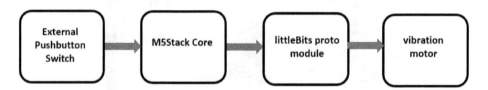

2. Interactive quiz 2: The vibration motor will initially be off.
3. Interactive quiz 3: The voltage source from the B port will be approximately 3.3V.
4. Interactive quiz 4: The digital toggle pin code block.
5. Interactive quiz 5: $Xc = 15.9\ K\Omega$.
6. Interactive quiz 6: Change the Wait code block value to 1 for 1 second.
7. Interactive quiz 7:

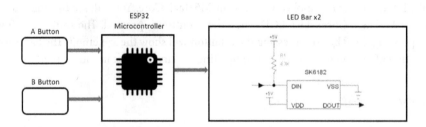

- Interactive quiz 8: The counter start value will be 2.

Part 2: M5Stack Electronic Interfacing Circuit Projects

The learning objective for *Part 2* is to instruct you on wiring electronic interfacing circuits to attach and operate SNAP circuit devices, discrete electronic breadboard circuits, littleBits electronic modules, and Arduino-based projects. To accomplish this goal, you will perform basic coding tasks using the UiFlow Arduino C/C++ software and wiring external SNAP circuits, discrete electronic breadboard circuits, and Arduino-based projects.

This part has the following chapters:

- *Chapter 4, It's a SNAP! Snap Circuits and the M5Stack*

- *Chapter 5, Solderless Breadboarding with the M5Stack*

- *Chapter 6, M5Stack and Arduino*

4

It's a SNAP! Snap Circuits and the M5Stack Core

In the previous chapter, you learned about the various hardware circuits that extend the operational functions of the **M5Stack Core**. You learned about the various hardware circuits that allow motion, light, and sound to appeal to the user of an M5Stack Core. You learned how to access the UiFlow Blockly code palette to program the M5Stack Core to utilize a vibration motor, a speaker, and the RGB LED. Furthermore, you learned how to wire a littleBits vibration motor to the M5Stack Core and program it to operate as a haptic device. With the knowledge of creating a block diagram and electronic circuit schematic diagram obtained in *Chapter 3*, you will be able to wire the M5Stack Core to operate various Snap Circuits devices. In this chapter, you will explore the M5Stack Core as a mini, programmable controller that operates various electronic circuits. You will be introduced to Snap Circuits, followed by an explanation of a version of an electronic learning product called the Arcade Games kit; you will learn how to enhance the Arcade Games user experience with Snap Circuits using an M5Stack Core; and you will learn how to wire the M5Stack Core using a basic electrical interfacing control circuit.

By the end of this chapter, you will have learned how to perform the following technical tasks with the M5Stack and Snap circuits:

- Drawing a typical M5Stack Core mini controller circuit schematic diagram
- Wiring and programming a Snap Circuits **LED Display and Microcontroller** (**LED MC**) module dice game
- Wiring, programming, and controlling an M5Stack Core Snap Circuits MC LED numbers and letters device
- Wiring, programming, and controlling an M5Stack Core Snap Circuits alarm device
- Wiring, programming, and controlling an M5Stack Core Snap Circuits counter with a sound device

In this chapter, we are going to cover the following main topics:

- What are Snap Circuits?
- What is a Snap LED MC module?
- Building an M5Stack Core Snap Circuits numbers and letters device
- Building an M5Stack Core Snap Circuits alarm device
- Building an M5Stack Core Snap Circuits counter with a sound device

Technical requirements

To engage with the chapter's learning content, you will need the **M5GO IoT Starter Kit** to explore the internal and external hardware, electronics, and software coding of your first basic application. You will need the **Snap Circuits Arcade Game learning product**, an electromechanical relay, and jumper wires. The UiFlow code block software will be required to build and run the M5Stack Core mini programmable controller applications.

You will require the following:

- The Snap Circuits Arcade Game learning product
- The M5Stack Core
- The M5GO IoT Starter Kit
- UiFlow code block software
- Snap Circuits jumper wires
- A USB C cable
- 3 AA batteries
- A Snap Circuits AC adapter
- An electromechanical relay module

The GitHub repository for the software resources is at `https://github.com/PacktPublishing/M5Stack-Electronic-Blueprints/tree/main/Chapter04`.

What are Snap Circuits?

Snap Circuits are electrical and electronic components mounted on colorful plastic shapes. Each colorful shape has a **metal snap** at key connecting points of the electrical and electronic component. The electrical and electronic component provides the specific **function** for the Snap Circuits part.

Therefore, a Snap Circuits part can be described as a functional block based on an electrical and electronic component mounted onto a plastic base. *Figure 4.1* shows some example Snap Circuits blocks:

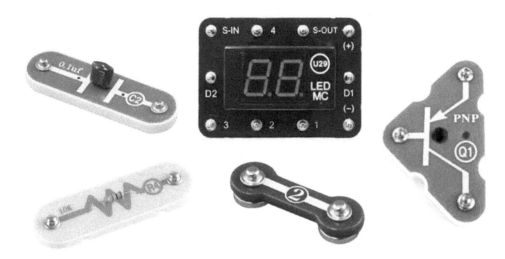

Figure 4.1 – Snap Circuits blocks

The Snap Circuits block designated by the number *2* is a wire block. You connect Snap Circuits blocks using this component. There are various types of wire blocks based on length. For example, there are wire blocks with lengths of 3, 4, and 5. The 1-wire snap is packaged into a plastic circle. You will use this snap component to provide an **electrical stand-off** to structurally align the other Snap Circuits and wire blocks properly. *Figure 4.2* shows a 1-snap wire:

Figure 4.2 – Snap Circuits blocks

Besides using wire blocks to attach Snap Circuits components together, you can use **jumper wires**. Like the wire blocks, jumper wires provide a flexible approach to connecting Snap Circuits blocks remotely. For example, a pushbutton switch Snap Circuits block can be wired several inches from the main electrical sound, control, or amplifying circuit. You can create alarm circuits that require trip wires to be strategically placed in a designated detection zone. Some examples of Snap Circuits jumper wires are shown here:

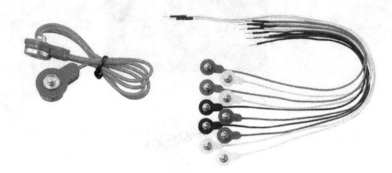

Figure 4.3 – Snap Circuits jumper wires

To provide a stable platform to mount and attach the Snap Circuits blocks, you will use a **base grid**. You may think of the base grid as a printed circuit for wiring Snap Circuits. The Snap Circuits, wire blocks, and jumper wires are placed onto the base grid plastic stubs. The placement of these components provides **mechanical rigidity** and proper **electrical conductivity** for the Snap Circuits device. *Figure 4.4* shows a Snap Circuits base grid:

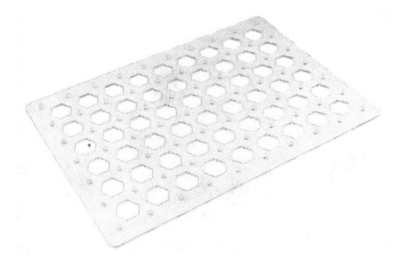

Figure 4.4 – Snap Circuits base grid

Now that you understand Snap Circuits, in the next section, you will learn about the **LED Display and Microcontroller** (**LED MC**) module.

What is a Snap LED MC module?

The Snap LED MC module is a **dual seven-segment LED display** driven by a **microcontroller** and supporting electronic parts. The microcontroller is a **PICAXE-08M2 integrated circuit**. The PICAXE-08M2 microcontroller is an eight-pin integrated circuit. The chip has **flash memory**, thus allowing it to be programmed easily. *Figure 4.5* shows a photo of a PICAXE-08M2 microcontroller:

Figure 4.5 – The PICAXE microcontroller

The LED MC is pre-programmed with 21 arcade-based games. Some of the game titles are binary coded decimal, changing speed, the baseball game, and home run derby. The arcade games are accessible using a selector switch and a **sequence of instructions**. Several snaps provide various electrical control functions that allow the LED MC to provide the arcade games previously listed. *Figure 4.6* shows the LED MC with its snap functions:

(+): Battery
(-): Battery ground
(S-IN): Selector (S8) input
(S-OUT): An output, typically connected to an LED
(1): An output, typically connected to an LED
(2): An output, typically connected to a speaker
(3): An input, Selector (S8) input
(4): An output, typically connected to an LED
(D1): Turns off the right LED display
(D2): Turns off the left LED display

Figure 4.6 – The LED MC

The circuit schematic diagram for the LED MC is shown in *Figure 4.7*:

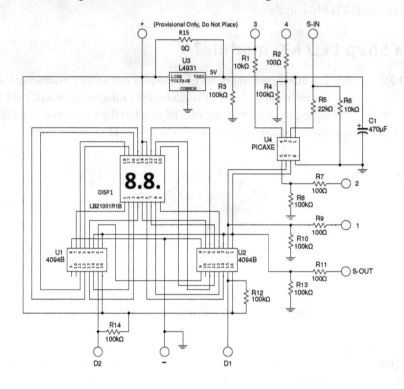

Figure 4.7 – The LED MC electronic circuit schematic diagram

With your knowledge about the LED MC, you will build a Snap Circuits arcade game numbers and letters device operated by an M5Stack Core. You will learn how to wire the numbers and letters device using Snap Circuits blocks and an M5Stack Core. You will learn how to wire an electromechanical relay to an M5Stack Core to operate the Snap Circuits device.

Building an M5Stack Core Snap Circuits numbers and letters device

The Snap Circuits Arcade Game numbers and letters device will allow a sequence series of letters and numbers to be displayed on the LED MC. Typically, the selector switch's **C** button is used to manually sequence the letters and numbers. You will **bypass** the selector switch with an M5Stack controller. The M5Stack controller will be wired to an electromechanical relay. The electromechanical relay's contact will act as the replacement for the **C** button. A basic block diagram of the controller design is shown in *Figure 4.8*:

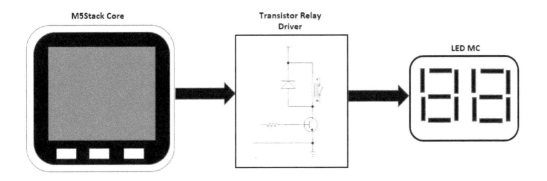

Figure 4.8 – M5Stack controller block diagram

You will use an off-the-shelf electromechanical relay module to act as the switching interface device between the M5Stack Core and the LED MC. This kind of electromechanical relay module device is shown in *Figure 4.9*. The electromechanical relay module has one set of normally opened switching contracts – therefore, the component is known as a **single-pole, single-throw** device. In datasheets and electrical engineering textbooks, the single-pole, single-throw electromechanical relay normally-opened contacts are abbreviated to **SPST NO**.

> ### Quiz 1
> If SPST NO is the abbreviation for single-pole, single-throw normally open, what is SPST NC?

Figure 4.9 – An electromechanical relay module with SPST NO contacts

The electronic circuit schematic diagram symbol for an electromechanical relay with SPST NO contacts is shown in *Figure 4.10*.

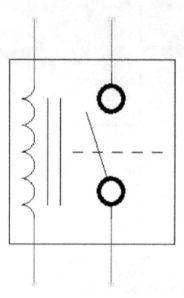

Figure 4.10 – A typical schematic diagram of an electromechanical relay with an SPST NO contacts symbol

The electromechanical relay module uses a **transistor** to drive or operate the coil of the electromechanical relay. The transistor will provide a ground for the electromechanical relay upon being biased properly by the resistor combination of R1 and R3. The combination of R1 and R3 forms a **voltage divider** circuit capable of operating the transistor properly. The **biasing** or the DC operating point will provide sufficient current flow from the transistor's **collector-emitter circuit**. With sufficient current flowing through the collector-emitter circuit, the transistor will provide a ground or low-side voltage rail for one side of the electromechanical relay's coil. The other side of the electromechanical relay's coil is wired to the **high side** or 3 V DC rail established by the 3 V **voltage regulator**.

The M5Stack Core control voltage provides the proper biasing for the transistor-relay driver circuit to operate the electromechanical relay. The M5Stack Core's control voltage is applied at point D1 of the R1 resistor. The **VCC** or **collector supply voltage** is a 5 V DC source provided by the M5Stack Core. The electronic circuit schematic diagram for the electromechanical relay module is shown in *Figure 4.11*:

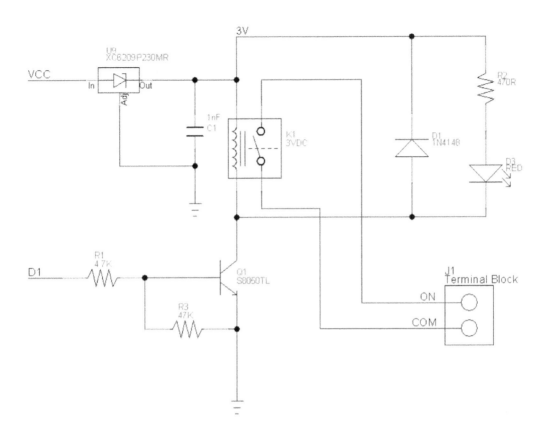

Figure 4.11 – An electronic circuit schematic diagram of the electromechanical relay module

You will connect the electromechanical relay module and the M5Stack Core using individual jumper wires. The individual wires are inserted into the three respective **cavities** of the jumper harness connector. The colors of the three respective cavities are red, black, and white. The electrical wiring diagram you will use to connect the M5Stack Core to the electromechanical relay module to build the M5Stack controller is shown in *Figure 4.12*:

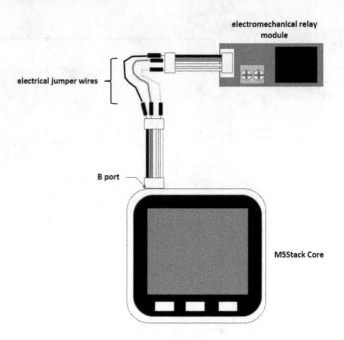

Figure 4.12 – M5Stack controller electrical wiring diagram

In addition to the electrical wiring diagram shown in *Figure 4.12*, you may use *Figure 4.13* to assist in attaching the M5Stack Core to the electromechanical relay module. You will notice in the figure that the electrical wiring scheme is equivalent to what is depicted in *Figure 4.12*. You will use this mini programmable controller to operate the Snap Circuits Arcade Games numbers and letters device.

Figure 4.13 – Pictorial of an M5Stack controller

> **Quiz 2**
>
> In reviewing the electrical wiring diagram shown in *Figure 4.12*, what function does the white jumper wire provide the M5Stack controller with?

With the electromechanical relay attached to the M5Stack Core, you are ready to build the Snap Circuit numbers and letters device. You will use the pictorial diagram in *Figure 4.14* to guide your building of the Snap Circuit numbers and letters device:

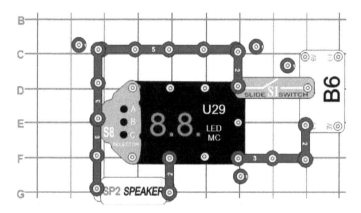

Figure 4.14 – Pictorial drawing of the Snap Circuits numbers and letters device

You will use snap number one (*1*) shown in *Figure 4.14* to properly align the snap wires appropriately to the matching electronic components. You will attach the electromechanical relay module's SPST NO contacts to the selector switch using Snap Circuits jumper wires. You will notice an AC adapter (**B6**) is used instead of a battery pack. The AC adapter is used to avoid purchasing batteries. You may use a battery pack or the AC adapter's 5 V source to power your Snap Circuits device. The jumper wires will be attached to points **E** and **F** on the **base grid**. *Figure 4.15* shows the Snap Circuits jumper wires attached to the electromechanical relay SPST NO contact:

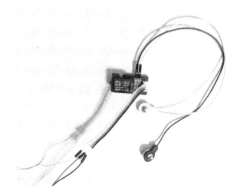

Figure 4.15 – Attaching Snap Circuits jumper wires to the electromechanical relay module

Quiz 3

Looking at *Figure 4.14*, the slide switch (**S1**) is attached to what snap of the LED MC?

A bigger-picture perspective of the M5Stack Core interfaced with the Snap Circuits numbers and letters can be viewed in the electronic circuit schematic diagram. You will notice that the electromechanical relay module's SPST NO contact is wired in parallel with the **C** switch of the selector switch. The **C** switch is bypassed using the electromechanical relay module's contact, operated by the M5Stack Core. *Figure 4.16* shows the electronic circuit schematic diagram:

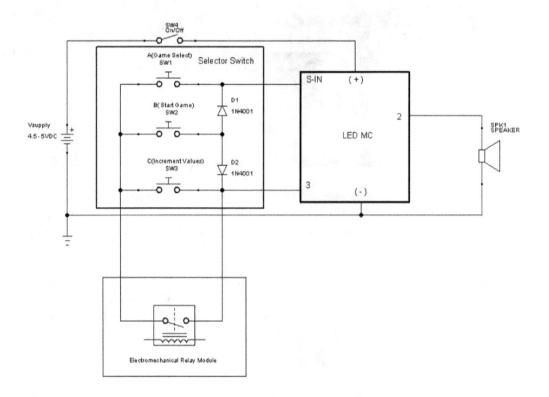

Figure 4.16 – Electronic circuit schematic diagram for the M5Stack
controller Snap Circuits numbers and letters device

You can use the following pictorial to assemble the final M5Stack controller Snap Circuits numbers and letters device:

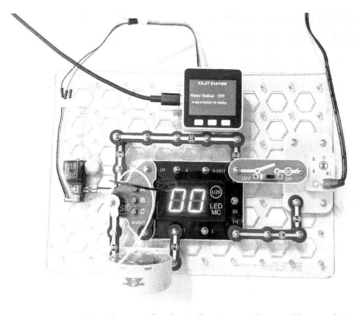

Figure 4.17 – M5Stack controller Snap Circuits number and letters device

To secure the electromechanical relay module, you can use two small globs of mounting putty. Place the mounting putty on each side and on top of the **Printed Circuit Board** (**PCB**) to the base grid. You can attach the M5Stack Core to the base grid using a small **LEGO plate** and **brick**. The LEGO plate and brick combination will attach to the back of the M5Stack Core. Placing mounting putty on the bottom of the LEGO brick will secure the M5Stack Core to the base grid. *Figure 4.18* illustrates this mechanical attachment scheme:

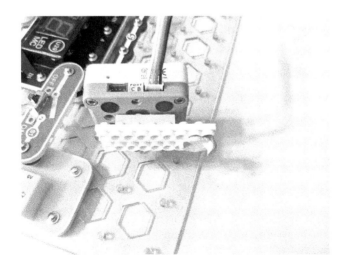

Figure 4.18 – Adding a mechanical attachment feature to the M5Stack Core

Congratulations, you have assembled the M5Stack controller Snap Circuits numbers and letters device! Pressing the **C** switch manually will sequence and display the numbers and letters on the LED MC module. In the next step of this project, you will add the control software to the M5Stack Core. The software will allow the Snap Circuits number and letters device to operate automatically. The M5Stack Core's software will allow the LED MC's program to sequence through a series of numbers and letters automatically. In total, there are 29 letters and numbers pre-programmed into the LED MC.

Game selection procedure

Before proceeding in developing the UiFlow Blockly code for the M5Stack Core, here is the LED MC game selection procedure you may use for obtaining the Snap Circuits numbers and letters arcade game. You will first turn on the LED MC by sliding the **S1 switch** to the on position. The LED MC will display **00**. You will press the **A** button on the **selector switch**. You will continue to press the switch until **11** is displayed. You will press the **B** button on the selector switch. The LED MC will display **Go** (the word) and the speaker will beep. Lastly, you will press the selector switch's **C** button to sequence the LED MC to display the 29 numbers and letters.

> **Interactive quiz 1**
>
> By changing the repeat value to 5, what number will be displayed on the LED MC?

The M5Stack Core controller's software

You will program the M5Stack Core's controller using UiFlow Blockly code. UiFlow Blockly code will provide control over the electrical wired electromechanical relay's SPST contact. You will be able to start the sequencing of the numbers and letters using the UiFlow Blockly code UI. With the UI, the sequence of the numbers and letters will be initiated by the electromechanical relay's SPST contact. *Figure 4.19* shows the M5Stack controller UiFlow Blockly code:

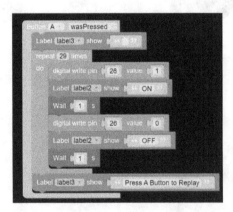

Figure 4.19 – The M5Stack controller UiFlow Blockly code

You will use the UI layout to define the text and the appearance of the M5Stack controller, as shown in *Figure 4.20*:

Figure 4.20 – The M5Stack controller UI

To execute the device, use the steps discussed in the game selection procedure. Remember, you will not be using the **C** button on the selector switch. Instead, the M5Stack controller will automatically be sequenced through the pre-programmed code of the LED MC. You will execute the UiFlow Blockly code to the M5Stack Core using the UiFlow **run button**. You now have a **development** template to construct and operate the remainder of the projects in this chapter.

Building an M5Stack Core Snap Circuits alarm device

In this section, you will wire a Snap Circuits alarm device that is controlled by the M5Stack controller. The alarm circuit device will be operated by the electromechanical relay module used in the previous project. The electromechanical relay module's switching action will be influenced by the M5Stack Core's UiFlow Blockly code. The UiFlow Blockly code will alternately switch the Snap Circuits alarm device on and off. The switching contact of the electromechanical relay module will provide a symmetric-based alarm sound, thus getting the attention of a bystander. The main Snap Circuits module that will provide the audible sound is the ALARM IC. *Figure 4.21* shows the Snap Circuits alarm IC and electrical connections.

> **Note**
> The Snap Circuits alarm IC module uses a custom **Application-Specific Integrated Circuit** (**ASIC**) to produce several audible sounds.

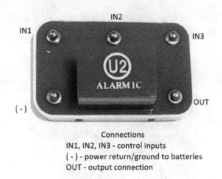

Connections
IN1, IN2, IN3 - control inputs
(-) - power return/ground to batteries
OUT - output connection

Figure 4.21 – The Snap Circuits ALARM IC module

The ALARM IC module uses an ASIC that allows various sounds to be produced. The sounds that can be produced with the electronic module are a siren, machine gun, fire engine, and European siren. To aid in the sound generation produced by the electronic module, there is a **resistor** and **NPN transistor** parts wired to the custom IC. The resistor provides sufficient gain for the **ALARM IC** module with the NPN transistor driving a speaker. The circuit schematic diagram of the custom IC design is shown in *Figure 4.22*.

> **Quiz 4**
>
> In reviewing *Figure 4.21*, a pushbutton switch would be attached to which connection snap of the **ALARM IC module**?

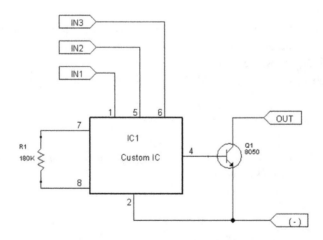

Figure 4.22 – The custom IC circuit schematic diagram

You can start to assemble the M5Stack controller Snap Circuits ALARM IC device using the pictorial diagram shown in *Figure 4.23*. *Figure 4.23* illustrates the placement locations of the ALARM IC module, M5Stack Core, and the electromechanical mechanical relay module on the base grid. With the placement of the components on the base grid, the device can easily be tested using the slide switch (**S1**).

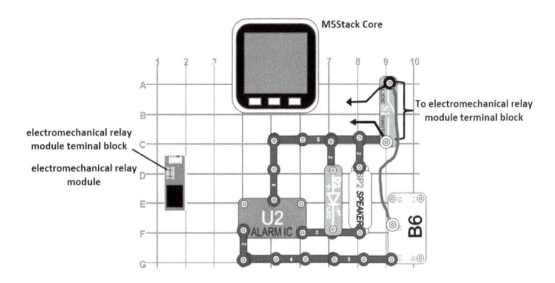

Figure 4.23 – Placement of the Alarm IC device components on the Snap Circuits base grid

Place the slide switch (**S1**) in the on position. With the slide switch in the on position, an alarm sound shall be heard through the speaker. Congratulations, you have built a Snap Circuits alarm device!

> **Note**
>
> You can design your own assembly layout diagrams using the Snap Circuits Designer. To obtain the design tool, go to the following website: `https://www.elenco.com/snap-circuits-designer/`.

As a reference, the electronic circuit schematic diagram for the Snap Circuits ALARM IC circuit is shown next:

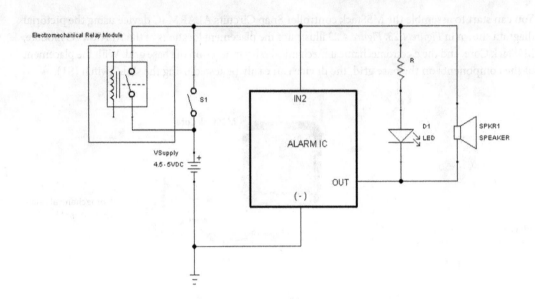

Figure 4.24 – Snap ALARM IC circuit schematic diagram

You will notice a series limiting resistor (**R**) and LED (**D1**) in the electronic circuit schematic shown in *Figure 4.24*. The D2 block is constructed using these two discrete electronic components. Therefore, you can wire your own LED circuit following the circuit configuration shown in *Figure 4.24*. A 330-ohm resistor value will provide the appropriate current limiting needed to prevent the LED from being damaged. As a final assembly reference, here is the completed project build of the M5Stack controller – the Snap Circuits ALARM circuit device is shown in *Figure 4.25*. You are now ready to program the M5Stack Core to operate the Snap Circuits ALARM device using the UiFlow Blockly code software.

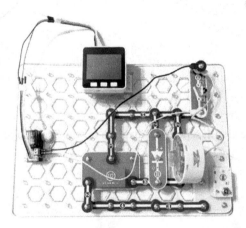

Figure 4.25 – The completed built M5Stack controller – Snap ALARM device

You can use the UI layout as shown in *Figure 4.26* to create the operator controls to operate the Snap Circuits ALARM device:

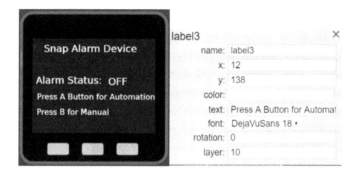

Figure 4.26 – The Snap ALARM device UI controls layout

You can use the following UiFlow Blockly code to operate the Snap Circuits ALARM device using the M5Stack Core:

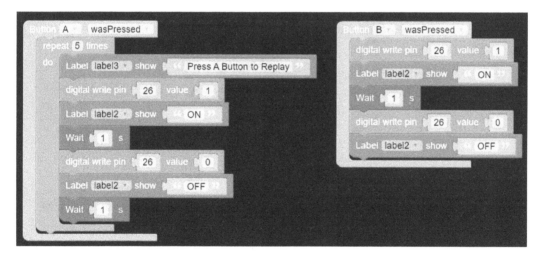

Figure 4.27 – M5Stack Core Blockly code

You will start the Snap Circuits alarm circuit operation by executing the code shown in *Figure 4.27* to the M5Stack Core using the UiFlow **run** button. To ensure the proper operation of the alarm under the control of the M5Stack Core, you will place the slide switch (**S1**) in the off position. Pressing the **A** button on the M5Stack Core will cause the alarm to sound repeatedly five times. The **B** button will cause the alarm to sound once for 1 second. *Figure 4.28* shows the M5Stack Core's UI in operation.

You can create a fire engine sound with your M5Stack controller Snap Circuits Alarm device by adding a snap wire between the **IN1** and **ground** (-) snaps. *Figure 4.29* illustrates this electrical connection. With this suggested modification, document the change in a journal or spiral notebook.

> **Note**
>
> As suggested in the previous paragraph, take a journal or spiral notebook and transform it into an engineer's notebook to document new M5Stack Core design ideas!

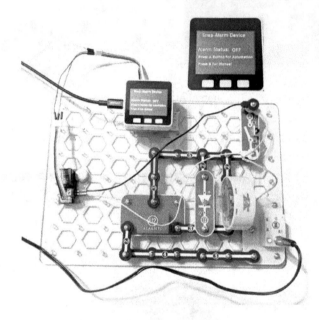

Figure 4.28 – A functional M5Stack Core UI

Creating a fire engine sound using this Snap Circuits modification design is shown in *Figure 4.29*:

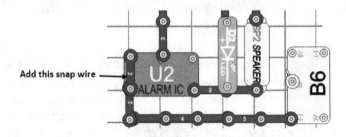

Figure 4.29 – Fire engine sound

> **Interactive quiz 2**
>
> Modify the Snap Circuits ALARM as shown in *Figure 4.30*. Run the Snap Circuits ALARM device code on the M5Stack Core: what sound is produced by the modification?

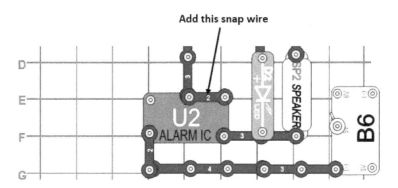

Figure 4.30 – Interactive quiz 2 circuit

Congratulations – you have successfully built an interactive alarm using Snap Circuits components and the M5Stack Core. As illustrated in this project, the M5Stack Core provides a control method for operating the alarm using a manual or automation UI approach. The M5Stack Core allows the alarm to be repetitive or single-operated by pressing the **operate** button on the ESP32-based programmable controller. In the final project, you will learn how to add sound with a decimal counter. You will use the LED MC, the alarm IC, and a speaker to create this unique analog and digital device.

Building an M5Stack Core Snap Circuits counter with a sound device

In this final project, you will use the knowledge obtained on building Snap Circuits to construct a counting sound device. You will learn how to repurpose the alarm device to work with the LED MC. The project's concept of reuse consists of controlling the counting feature of the LED MC and adding an audible sound. The audible sound resembles a machine gun when the ALARM IC module is triggered. The LED MC will rapidly display numbers based on the module's S-IN input being triggered for a specified time duration. You will make a slight modification to the M5Stack Core's UI to reflect the name of the project. The concept block diagram for the M5Stack controller counting sound device is shown in *Figure 4.31*:

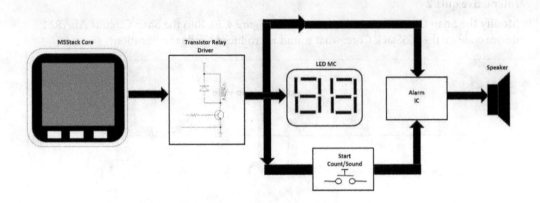

Figure 4.31 – M5Stack controller Snap Circuits counting sound device block diagram

Reviewing the block diagram shown in *Figure 4.31*, you will notice the **transistor relay driver circuit** triggering the LED MC and alarm IC. The length of the time for which the trigger signal is engaged is directed by the M5Stack Core's Blockly code. A long trigger signal time allows a rapid count to be displayed on the LED MC. In parallel with the transistor relay driver circuit, there is a manual **momentary** pushbutton switch capable of providing a similar trigger signal function. In contrast, the manual pushbutton switch can be considered a testing feature for the counting sound device. A long press of the momentary pushbutton switch will allow for rapid counting and a long duration of the machine gun sound.

To provide a circuit perspective view, an electronic circuit schematic diagram is shown in *Figure 4.32*. You will notice the electromechanical relay module wired in parallel with the momentary pushbutton switch. As discussed in the previous paragraph, you can test the counting sound device manually using the momentary pushbutton switch. The M5Stack Core Blockly code provides both automated and manual modes of operation for the counting sound device.

> **Note**
> Connecting the **IN2** and **IN3** snap pins will allow the alarm IC module to produce a machine gun sound.

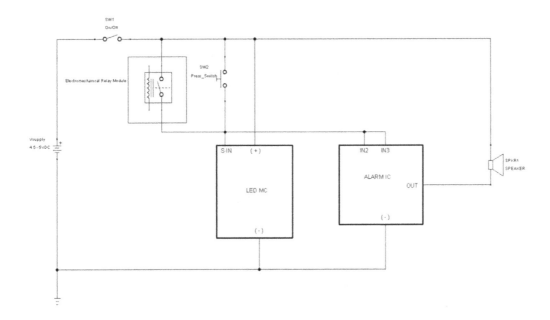

Figure 4.32 – M5Stack controller – electronic circuit schematic diagram of the counting sound device

You will use a pictorial diagram to electrically wire the electrical electronic components of the M5Stack controller counting sound device. The pictorial diagram to aid you in the wiring of the device is shown in *Figure 4.33*. The unique design of the circuit means that the LED MC does not provide a trigger control signal to the ALARM IC module. The trigger control signal is provided by the momentary pushbutton switch or the electromechanical relay module contacts.

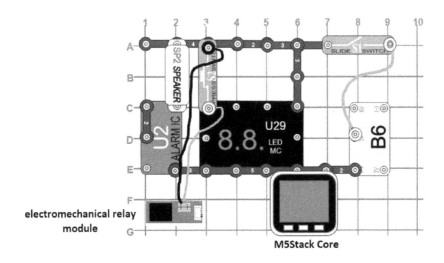

Figure 4.33 – Pictorial wiring diagram for the counting sound device

Quiz 5

In digital electronics, there is an electrical property of electromechanical switches that is disruptive to counting or sequential circuits but provides a unique aesthetic appeal to the LED MC. What is the name of this disruptive electromechanical switch property?

Once you have assembled the M5Stack controller counting sound device, your final project should resemble the picture shown in *Figure 4.34*:

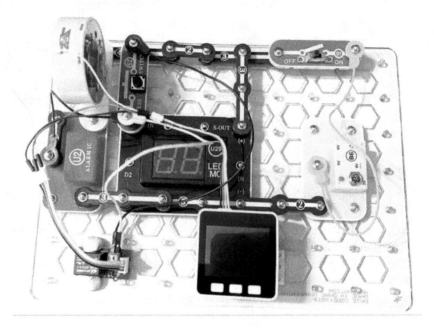

Figure 4.34 – Final project build of the M5Stack controller counting sound device

As previously discussed in this chapter, mounting putty is used to secure the electromechanical relay module to the Snap Circuits base grid. The M5Stack Core is attached to the Snap Circuits base grid using a LEGO brick and plate. The LEGO brick plate unit is attached to the M5Stack Core. You will attach the M5Stack Core LEGO brick plate unit to the Snap Circuits base grid using mounting putty. Refer to *Figure 4.18* for the placement of the mounting putty on the LEGO plate.

The final step to complete the M5Stack controller counting sound device is the UiFlow Blockly software. You will use the alarm circuit software as is, except for a small change to the M5Stack Core UI. The software heading is renamed as `Counting Sound Device`. The change in UI is shown in *Figure 4.35*:

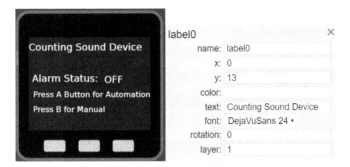

Figure 4.35 – Counting Sound Device UI

The UiFlow Blockly code has the same function as for the alarm circuit device project. Therefore, the Blockly code is presented here for convenience:

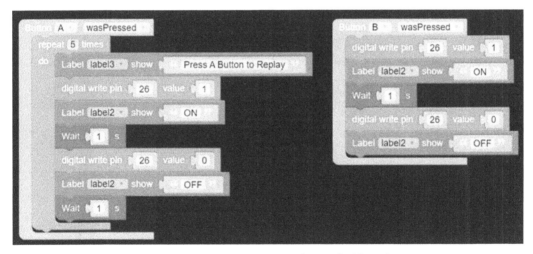

Figure 4.36 – The ALARM circuit device Blockly code

You will press the **run** button on the UiFlow taskbar to execute the Blockly code on the M5Stack Core. You will see the **Counting Sound Device** name displayed on the M5Stack Core's TFT LCD. *Figure 4.37* illustrates the functional M5Stack controller counting sound device. Congratulations, you have successfully built an interactive Snap Circuits counting sound device.

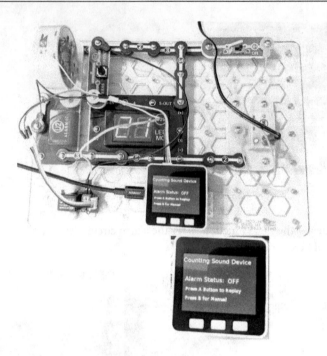

Figure 4.37 – Functional M5Stack controller counting sound device

> **Interactive quiz 3**
>
> In reviewing *Figure 4.36*, true or false – changing the **Wait** instruction time to 500 ms for the **Button A wasPressed** code blocks will slow down the LED MC counting sequence?

Congratulations – you have completed the hands-on activities and interactive quizzes in this chapter!

Summary

In the chapter, you learned about the electronic circuits used in the Snap Circuits Arcade Games kit. You learned how to wire electronic circuits using the Snap Circuits modules. You learned about the PICAXE 8-bit microcontroller and its application with the LED MC. You learned that the ALARM IC module has an ASIC. You learned about the workings of the ASIC by building an alarm circuit device. You also learned how to wire an electromechanical relay module to the M5Stack Core. You learned about the technical specifics of the block diagram and how it aligns with an electronic circuit schematic diagram. With this knowledge, you learned how to wire the M5Stack Core to the specific Snap Circuits project using an electromechanical relay.

Furthermore, you learned how to make a programmable controller using the M5Stack Core to operate the LED MC and the ALARM IC module. The LED MC's Picaxe microcontroller software operates the numbers and letters game. The LED MC's numbers and letters game is enabled by the M5Stack Core driving the electromechanical relay module's SPST NO contact. The enabling feature of operating the electromechanical relay module was directed by the M5Stack Core UiFlow Blockly code's code blocks. The enhancements to the Snap Circuits projects were accomplished using the M5Stack Core programming capabilities to operate port B, which switched the wired electromechanical relay module on and off. In addition, you used your coding knowledge to answer the interactive quizzes through hands-on investigation. Lastly, you explored the repurpose or remix prototyping concept by modifying the alarm circuit code to operate the counting sound device.

In this next chapter, you will take this knowledge of building an M5Stack programmable controller to operate discrete solderless breadboard electronic circuits, a littleBits LED flasher module, and a temperature sensor.

Interactive quiz answers

Quiz 1: Single-Pole, Single-Throw, Normally Closed.

Quiz 2: The white jumper wire provides the control signal to operate the transistor relay driver circuit for the electromechanical relay module.

Quiz 3: The (+) snap pin.

Quiz 4: Switch contact bounce.

Quiz 5: The number 4.

Interactive quiz 1: A machine gun sound.

Interactive quiz 2: False: The 500ms wait instruction will speed up the LED MC counting sequence.

Solderless Breadboarding with the M5Stack

In the previous chapter, you learned about the various Snap Circuits blocks that extend the **M5Stack Core** operational functions. You learned about blocks that allow motion, lights, and sound to enhance the appeal to the user of an M5Stack Core. You learned how to program the LED MC using a basic procedure. The **PICAXE microcontroller** used in the LED MC was presented to you using an electronic circuit schematic diagram. Furthermore, you learned about the ALARM IC module and the ASIC used to produce various sound effects. With the alarm IC, you learned how to wire the Snap Circuits unit to produce sounds such as a machine gun, a siren, a fire engine, and a European siren. You learned about the snap pins for the LED MC and the alarm IC modules in the previous chapter. Lastly, an interface circuit technique for wiring the M5Stack Core to a Snap Circuits LED MC and alarm IC using a transistor relay driver circuit was presented in the previous chapter.

In this chapter, you will explore the transistor relay driver and additional interface circuits. With the knowledge of how to create a block diagram and electronic circuit schematic diagram obtained in *Chapter 3*, you will be able to wire the M5Stack Core to operate various electronic circuit devices. You will be introduced to littleBits electronic modules by wiring them to the M5Stack Core. Furthermore, you will learn how to wire a littleBits temperature sensor to the M5Stack Core and create a simulator monitor. A discrete LED electronic flasher and a transistor DC motor driver circuit controlled by an M5Stack Core will be presented in this chapter.

By the end of this chapter, you will know how to perform the following technical tasks with the M5Stack and solderless breadboarding:

- Build a Tinkercad Circuits transistor LED driver circuit model
- Wire and test an M5Stack Core discrete LED electronic flasher
- Wire and test an M5Stack Core discrete transistor motor driver circuit
- Wire and test an M5Stack Core littleBits temperature sensor simulator device
- Wire and test an M5Stack Core controller littleBits LED flasher

In this chapter, we are going to cover the following main topics:

- An introduction to electronic interfacing circuits

- Building a discrete transistor LED electronic flasher

- Building a discrete transistor DC motor driver circuit

- Building a littleBits temperature sensor M5Stack Core simulator monitor

- Building an M5Stack Core controller littleBits LED flasher

Technical requirements

To engage with the chapter's learning content, you will need the **M5GO IoT starter kit** to explore the internal/external hardware electronics and software coding of your first basic application. **littleBits**, **electronic modules**, discrete resistors, transistors, an LED, an electromechanical relay, and jumper wires will be required to complete the projects in the chapter. Furthermore, the UiFlow code block software will be required to build and run the M5Stack Core mini programmable controller applications.

You will require the following:

- A littleBits w9 proto module

- A littleBits i12 temperature sensor

- A littleBits o1 LED

- A littleBits o2 long LED

- A littleBits o25 DC motor or equivalent

- A USB-C cable

- A 2N3904 NPN transistor

- A 1N4001 silicon diode

- An electromechanical relay module

- An electromechanical relay 3-5 V DC

- Two 560-ohm, +/5% (green, blue, brown, gold) resistors

- A 10-Kiloohm, +/5% (brown, black, orange, gold) resistor

- A discrete LED

Here is the GitHub repository for the software resources:

```
https://github.com/PacktPublishing/M5Stack-Electronic-Blueprints/
tree/main/Chapter05
```

An introduction to electronic interfacing circuits

Wiring two non-compatible circuits requires an electrical connection method. The electrical connection method addresses control signal differences between the two non-compatible circuits. **Electronic interfacing** is an electrical wiring method that connects or links one circuit to another electrical device. Traditionally, there are two approaches to interfacing electronic circuits: **input** and **output**. Input interfacing is an electrical approach to adapting an electronic circuit's output to the input of another circuit device. Output interfacing allows an electronic or electrical circuit to control a **real-world device**. A real-world device such as an electromechanical relay, a discrete LED, or a **liquid-crystal display** (**LCD**) is wired to the output of a controller or switching circuit. The input and output electronic interfacing circuits traditionally use basic components, such as **resistors**, **capacitors**, or **transistors**. The arrangement of these basic components will allow either an amplification, filtration, or voltage-level translation to occur. *Figure 5.1* illustrates the concepts of basic input electronic interfacing circuits.

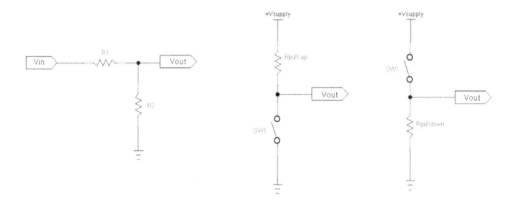

Figure 5.1 – Basic input electronic interfacing circuits

The idea behind these three basic input electronic interfacing circuits is to provide an **output voltage** (**Vout**) capable of allowing a compatible control signal to properly operate an **input port** or **pin** of a **microcontroller** or **digital circuit**. The basic input electronic interfacing circuits operate in the **digital domain**, thus providing a 0 or +**Vsupply** voltage to the input port or pin of a microcontroller or digital circuit. The voltage value of +**Vsupply** can be 3.3 V or 5 V, which is compliant with the input port or pin of the microcontroller or digital circuit. The two output electronic interfacing circuits are shown in *Figure 5.2*.

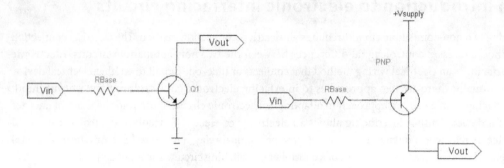

Figure 5.2 – Basic output electronic interfacing circuits

The transistors shown in *Figure 5.2* provide sufficient **amplification gain**, thus capable of **sinking** and **sourcing** the current flow needed to properly work with the output port or pin of the microcontroller or digital logic gate. Sinking the current flow allows the transistor to provide a switched ground to the wired electrical or **electromechanical load**. Sourcing the current provides an **active high-side voltage rail** to the wired electrical or electromechanical load. Traditionally, the output electronic interfacing circuit will operate an electromechanical load such as an electric motor or relay. Operating a **discrete** LED or **seven-segment LED display** requires a transistor or an **array** of transistors to turn the individual segments of the **optoelectronic** component on or off.

Besides using discrete switching components such as a transistor, an **IC** may be used for an electronic interfacing circuit. A buffer is an electronic circuit used to isolate the input from the output. The input and output are decided by default in relation to the electronic circuits wired together. The buffer's input **impedance** is high, therefore drawing little current flow to avoid disturbing the input of the target circuit. Impedance is the internal **AC resistance** of a circuit. *Figure 5.3* illustrates a typical buffer IC.

> **Quiz 1**
> Electronic interfacing allows two approaches: jumper wires and terminal blocks. True or false?

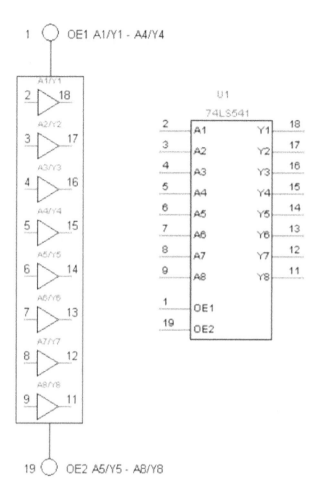

Figure 5.3 – A 74LS541 octal three-state buffer IC

You now have general knowledge about electronic interfacing and various approaches to wiring dissimilar circuits together. With this technical knowledge, you will be able to interface discrete circuits and littleBits electronic modules to the M5Stack Core. You will apply this technical knowledge in the next section by building a discrete transistor LED electronic flasher operated by an M5Stack Core.

Building a discrete transistor LED electronic flasher

You learned about electronic interfacing circuits in the previous section. In this section, you will apply this technical knowledge by building a discrete transistor LED electronic flasher operated by an M5Stack Core. To obtain the baseline information about the M5Stack Core-operated device, you will build a virtual transistor LED driver circuit. You will create the virtual transistor LED driver circuit using **Tinkercad Circuits**. Tinkercad Circuits is an online electronic circuit simulator platform that allows various discrete electrical electronic, digital, and microcontroller circuits to be constructed and tested. This online electronics laboratory was developed by Autodesk. You will use the electronic circuit schematic provided in *Figure 5.4* to build the Tinkercad Circuits transistor LED driver circuit.

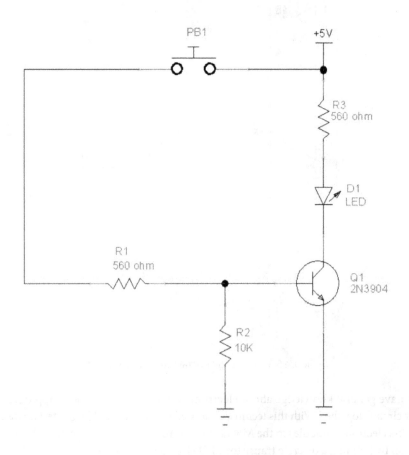

Figure 5.4 – The transistor LED driver circuit

Tinkercad Circuits is a popular online electronics **simulator** that allows you to test microcontrollers and digital and analog circuits. With Tinkercad Circuits, you can easily test the **electrical behavior** or **operation** of the transistor LED driver circuit shown in *Figure 5.4*. You can assess the Tinkercad Circuits online electronics simulator environment at this URL:

```
https://www.tinkercad.com/learn/circuits
```

After signing up and logging in to the free simulator website, you will see the Tinkercad Circuits **virtual lab bench** and **user interface** (**UI**) as shown in *Figure 5.5*.

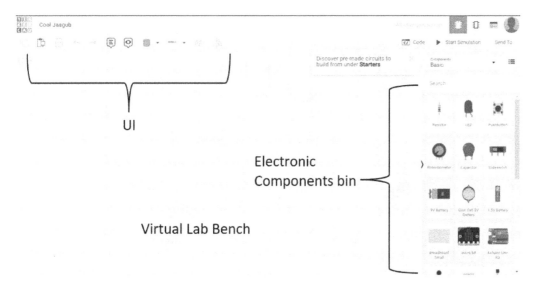

Figure 5.5 – The Tinkercad Circuits electronic simulator environment

You will create the circuit using a solderless breadboard and the electronic circuit components as shown in *Figure 5.4*. You can wire the transistor LED driver circuit using the **virtual model** as shown in *Figure 5.6*. You will adjust the DC power supply for a +5 V DC voltage value. Pressing the tactile pushbutton switch will allow the NPN **bipolar junction transistor** (**BJT**) to **sink current** or provide **ground** through the **collector-emitter** leads. The current flowing through the NPN BJT collector-emitter leads will turn on the LED.

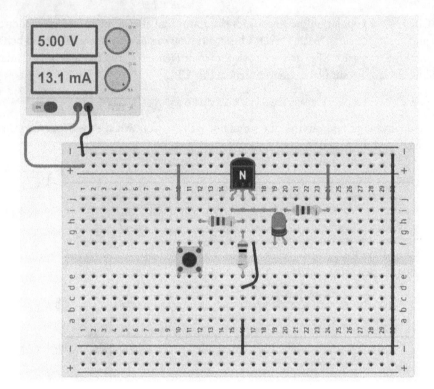

Figure 5.6 – A solderless breadboard transistor LED driver circuit

Figure 5.7 provides additional component value information to assist you in the electrical wiring of the transistor LED driver circuit. The reference designators are included to align the electronic circuit schematic diagram shown in *Figure 5.4* with the components shown on the solderless breadboard.

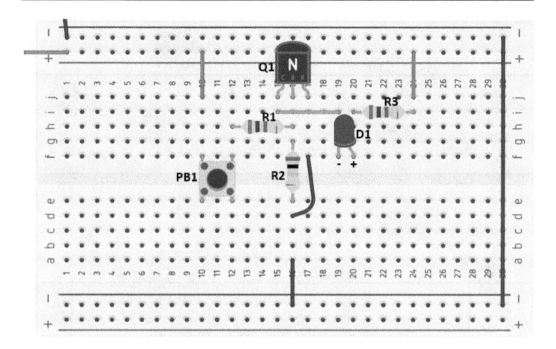

Figure 5.7 – Transistor LED driver circuit reference designators

You can add digital ammeters and a voltmeter to obtain additional electrical information regarding the driver circuit's operation. *Figure 5.8* shows the current branches of the transistor driver's base emitter and collector emitter. You will notice the voltmeter reading of 739 mV at the base of the transistor. The value displayed on the digital voltmeter is the **base-emitter junction voltage** (V_{BE}). Typically, for silicon transistors, the V_{BE} junction voltage is around 600 to 700 mV.

> **Note**
>
> To learn more about Tinkercad Circuits, refer to the online official guide at the following URL: `https://www.tinkercad.com/blog/official-guide-to-tinkercad-circuits`.

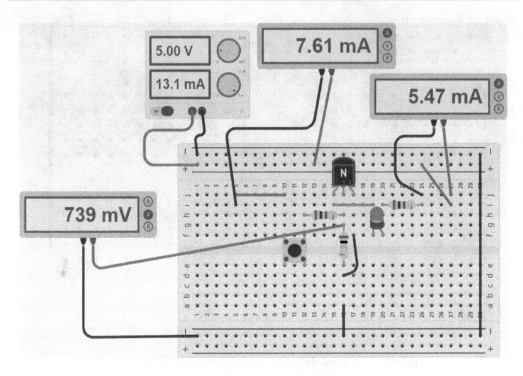

Figure 5.8 – Transistor LED driver circuit branch currents and V_{BE}

Quiz 2

Reviewing the DC power supply's ammeter reading shown in *Figure 5.8*, are the two branch currents correct?

Congratulations – you have successfully built and tested a discrete transistor LED driver circuit using the online Tinkercad Circuits simulator platform. You now understand the transistor LED driver's circuit operation and wiring configuration. With this knowledge, you will build a physical circuit using real electronic components and the M5Stack Core. The M5Stack Core interfacing concept is shown in *Figure 5.9*.

Note

The transistor shown in *Figure 5.8* has a pin orientation of **Collector (C)**, **Base (B)**, and **Emitter (E)**. The **2N3904 NPN transistor** has a pin orientation of **EBC**.

With the use of the circuit schematic and electrical wiring diagrams, you can wire up a physical transistor LED driver circuit prototype. The physical transistor LED driver circuit prototype can be wired to an M5Stack Core. You can program the M5Stack Core to operate the transistor LED driver circuit. Therefore, you can program the M5Stack Core to turn the transistor LED driver circuit on and off.

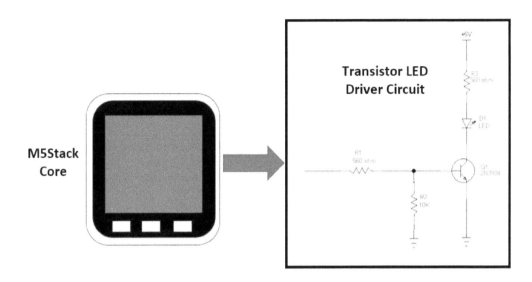

Figure 5.9 – M5Stack Core interfacing concept

With the M5Stack Core interfacing concept illustrated in *Figure 5.9*, along with the electrical wiring diagram shown in *Figure 5.10*, you can wire a transistor LED electronic flasher. The unique feature of this transistor LED electronic flasher is the ability to change the flashing rate of the optoelectronic-based circuit using this interfacing concept. The transistor LED electronic flasher wiring diagram is shown next.

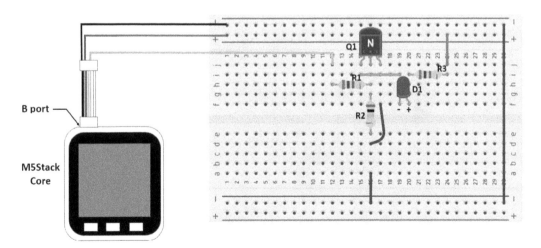

Figure 5.10 – Electrical wiring diagram for the transistor LED electronic flasher

You can also use the electronic circuit schematic diagram shown in *Figure 5.11* to wire the M5Stack Core-based transistor LED flasher device.

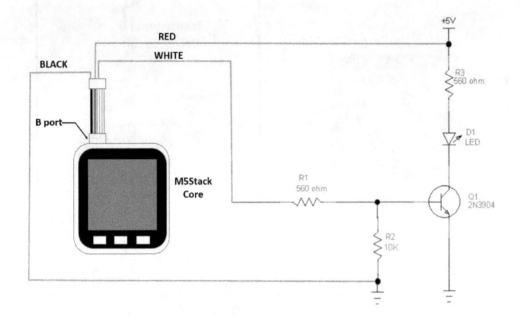

Figure 5.11 – Electronic circuit schematic diagram for the transistor LED electronic flasher

The author's completed wired transistor LED electronic flasher circuit prototype is shown next.

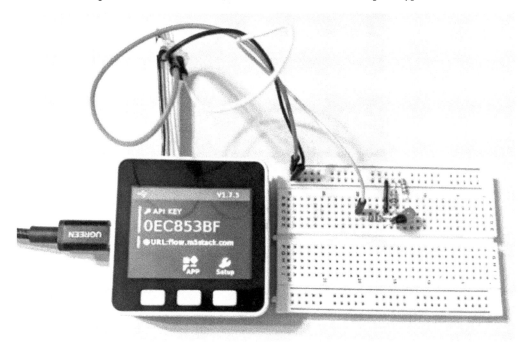

Figure 5.12 – Transistor LED electronic flasher circuit prototype

To operate the transistor LED electronic flasher, you will use the UiFlow Blockly code shown in *Figure 5.13*.

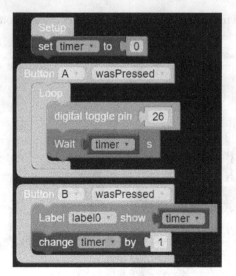

Figure 5.13 – UiFlow Blockly code for the programmable transistor LED flasher

Here, you can see the **label0** Blockly code in the **Label** code block assigned to the **timer** variable. You will provide a **UI** for the M5Stack Core to display a timer value when the **B** button is pressed. *Figure 5.14* shows the UI layout for displaying the timer value on the M5Stack Core LCD.

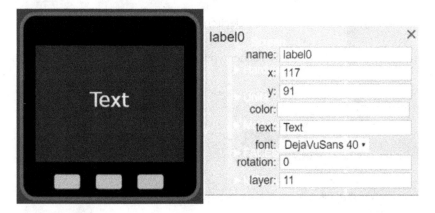

Figure 5.14 – Timer value UI layout

Interactive quiz 1

By modifying the **change timer** code instruction shown in *Figure 5.13* to **by** 2, what number will be displayed on the M5Stack Core LCD when running the code?

Upon running the code on the M5Stack Core, pressing the **B** button allows you to set the timer value. The timer value provides the on-and-off delays that allow the LED to blink. You will press the **A** button to allow the LED to blink at the set time delay value programmed using the **B** button. *Figure 5.15* illustrates setting the **timer** value by using the **B** button. To enter a new time delay value, double-press the M5Stack Core reset button and run the code on the ESP32-based device.

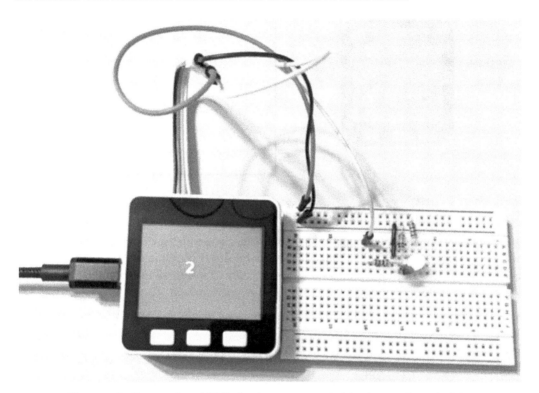

Figure 5.15 – The transistor LED flasher is programmed with a 2-second toggle delay

Congratulations – you have successfully wired and tested a transistor LED flasher that is enabled by the M5Stack Core. You used a transistor as an electronic switch to interface the M5Stack Core with a discrete LED. In the next section, you operate a small DC motor using the M5Stack Core.

Building a discrete transistor DC motor driver circuit

You have built and tested a transistor LED flasher using discrete electronic components. You have used the M5Stack Core as a programmable controller, thus setting various LED flash rates. You will use the operational concepts implemented previously with the transistor LED driver project to build an M5Stack Core-operated DC motor controller. *Figure 5.16* shows the concept of the M5Stack Core interfacing circuit.

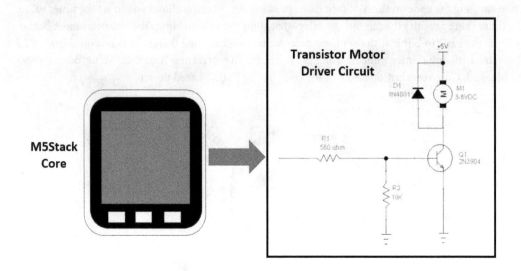

Figure 5.16 – The concept of the M5Stack Core interfacing circuit

With the M5Stack Core interface illustrated in *Figure 5.16*, along with the electrical wiring diagram shown in *Figure 5.17*, you can wire a transistor DC motor driver circuit to an M5Stack Core. The unique feature of this transistor DC motor driver is its ability to operate an electric motor using this interface. The transistor DC motor driver wiring diagram is shown next.

Interactive quiz 2

Redraw the transistor motor driver circuit using a PNP transistor.

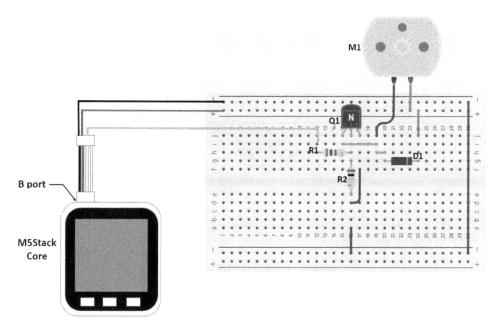

Figure 5.17 – Electrical wiring diagram of a transistor DC motor driver

You can also use the electronic circuit schematic diagram shown in *Figure 5.18* to wire the M5Stack Core-based transistor DC motor device.

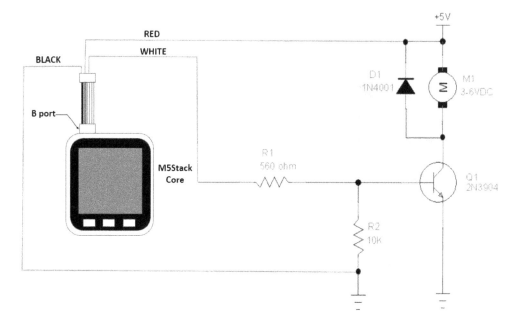

Figure 5.18 – Electronic circuit schematic diagram for a transistor DC motor driver

The author's completed wired transistor DC motor driver circuit prototype is shown next.

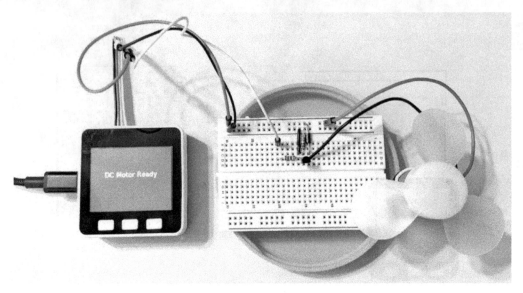

Figure 5.19 – Transistor DC motor driver circuit prototype

To test the transistor DC motor driver, you will use the UiFlow Blockly code shown in *Figure 5.20*.

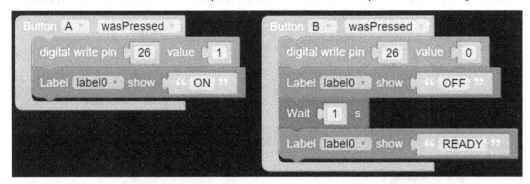

Figure 5.20 – UiFlow Blockly code for the M5Stack Core-enabled transistor DC motor driver

You will use a UI that is aesthetically appealing and informative for the user of the transistor DC motor driver circuit. The aesthetics consist of the controller's name displayed at the top, along with a **READY** prompt. Pressing the **A** button will start the **transistor driver** operating the DC motor. Pressing the **B** button will stop the DC motor. After a 1-second delay, the **READY** prompt is displayed. The UI that aligns with the Blockly code shown in *Figure 5.21* is shown next. The Blockly code will create a basic **on/off control** for the transistor DC motor driver.

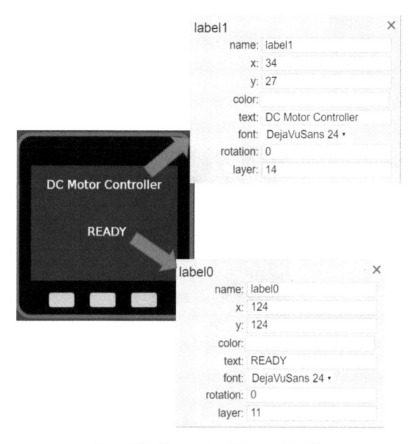

Figure 5.21 – The transistor DC motor driver UI

You can improve the UI layout shown in *Figure 5.21* by adding **ON** and **OFF** labels above the respective buttons. With this information included, there will be no confusion as to how to start and stop the DC motor. You have now developed a **human input module** (**HIM**) that provides direct control of the transistor DC motor driver circuit with text to operate the device. An HIM is an electronic module that allows you to configure the operation of the electric motor drive system. *Figure 5.22* shows the revised HIM-based UI.

Interactive quiz 3

If an LED with a limiting resistor is wired across R2, will the device turn on when the M5Stack Core's **A** button is pressed?

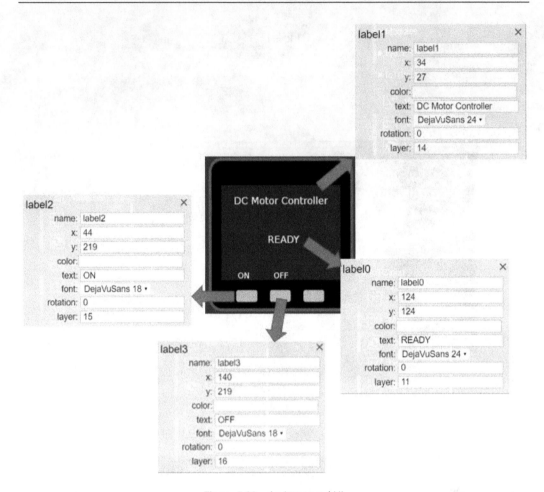

Figure 5.22 – An improved UI

Upon running the Blockly code shown in *Figure 5.20*, the improved UI will appear on the M5Stack Core's **thin film transistor** (**TFT**) LCD. *Figure 5.23* illustrates the actual improved M5Stack Core's UI. You now have an HIM-based UI to operate the transistor motor driver circuit easily without confusion.

Quiz 3

Which label attribute allows you to make the text red?

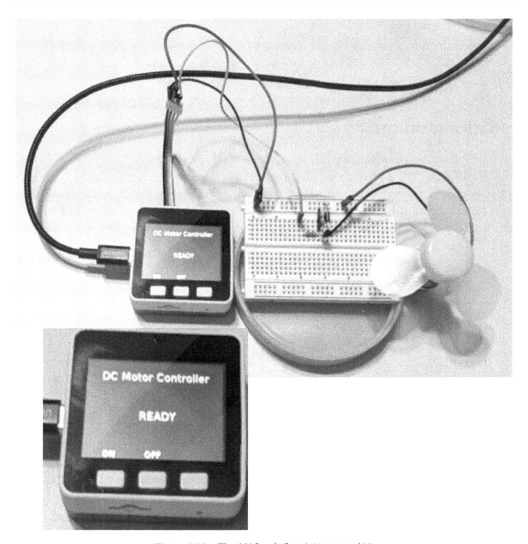

Figure 5.23 – The M5Stack Core's improved UI

Congratulations – you have completed the building and testing of a discrete transistor motor driver circuit. You have explored various on/off and timing controls to operate a discrete transistor motor driver circuit with an M5Stack Core. Moreover, you enhanced the UI by adding additional labels to make operating a wired DC motor with the M5Stack Core using a transistor driver circuit easier. In the next section, you will gain knowledge of **surface-mount circuits** using a **littleBits electronic module**. You will learn how to interface the littleBits electronic module to the M5Stack Core using a proto module to build a temperature sensor simulator module.

Interactive quiz 4

On the UI shown in *Figure 5.23*, can the **ON** and **OFF** text be changed to **START** and **STOP**?

Building a littleBits temperature sensor M5Stack Core simulator monitor

In this section, you will take your knowledge of the littleBits proto module and build a physical simulator with a monitor. The technical concept behind this simulator monitor project is to create a **physical model** of a temperature sensor using a littleBits slide dimmer control. By adjusting the slide dimmer control, a simulated temperature reading will be displayed on the M5Stack Core's **TFT LCD**. Upon the slide dimmer control reading exceeding a preset value, the M5Stack Core will turn on the internal LED bar. *Figure 5.24* shows the temperature sensor simulator monitor concept block diagram.

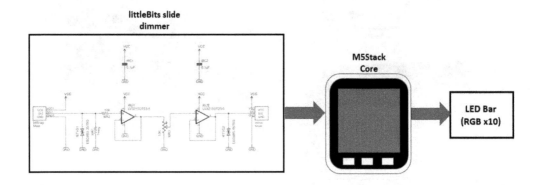

Figure 5.24 – Block diagram for the M5Stack Core temperature sensor simulator monitor

You will use the **angle unit** Blockly code block to capture the analog voltage values generated by the slide dimmer module. The angle unit Blockly code block will be aided by the label code block. The label code block will display the slide dimmer control value on the M5Stack Core's TFT LCD. The angle unit Blockly code block is displayed in *Figure 5.25*.

Figure 5.25 – The angle unit Blockly code block

You will need to add this unit to the palette of Blockly code blocks. With your mouse, click on the + plus sign to display the available M5Stack Core units. You will select the angle unit with your mouse. Once you have made the selection with a mouse click, the angle unit will be displayed on the UiFlow layout section of the UiFlow coding environment. *Figure 5.26* illustrates the placement of the angle unit in the UiFlow coding environment.

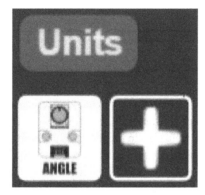

Figure 5.26 – Added angle unit to UiFlow layout area

You will use a slide dimmer control instead of the M5Stack Core's angle unit to simulate the operation of a temperature. A traditional temperature sensor reacts to various levels of hot and cold stimuli by changing its internal resistance. If the surrounding temperature is cold, the resistance will be high. Conversely, if the temperature level is high, the resistance will be low. The inverse behavior of the sensor reactions to temperature levels is known as **negative compensation**. To be more accurate, the inverse temperature-resistance sensor reaction is known as the **negative temperature coefficient** (**NTC**). A **thermistor** is a small resistive electrical component whose behavior to temperature versus resistance is based on the NTC operation. *Figure 5.27* illustrates an NTC thermistor and its temperature-to-resistance curve.

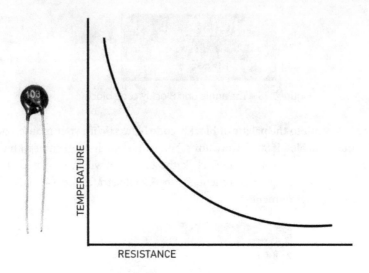

Figure 5.27 – A typical NTC thermistor's temperature-to-resistance curve

You will notice that as temperature increases, the thermistor's resistance decreases. With a decrease in temperature, the NTC thermistor's resistance increases. The simulator works in the same manner as an NTC thermistor. With the slide dimmer in the far-right position, the M5Stack Core's TFT LCD will display 0°C. As you move the slide dimmer to the left, the M5Stack Core's TFT LCD will show an increase in temperature. You now understand the temperature sensor simulator operation. You will begin the process of building the temperature sensor simulator shortly.

The first step in building the temperature sensor simulator is to wire an M5Stack Core jumper harness to a littleBits proto module. You will use *Figure 5.28* to wire the jumper harness to the proto module. The actual assembly build of the jumper harness to the proto module is shown in *Figure 5.29*. The M5Stack Core will power the slide dimmer control's **electronic interface circuit**. The red and black wires provide +5 V and ground to operate the two **operational amplifiers** soldered to the bottom side of the slide dimmer control.

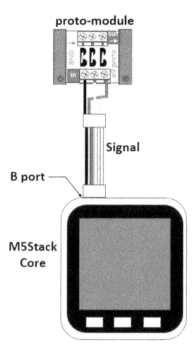

Figure 5.28 – Attaching the jumper harness to the proto module

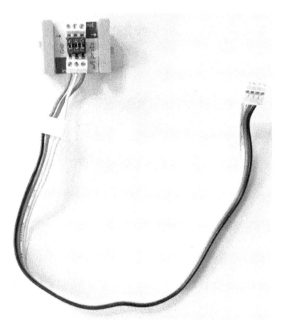

Figure 5.29 – Complete jumper wire harness assembly

The two operational amplifiers provide the appropriate **sourcing output current** and voltage required for the M5Stack Core to read the adjusted **signal values** produced by the slide dimmer control's potentiometer. *Figure 5.30* shows the electronic interface circuit of the slider dimmer control. The **bit snaps** on each end of the slide dimmer control provide the +5 V, GND, and signal required for the **electrical interface** with the M5Stack Core's B port.

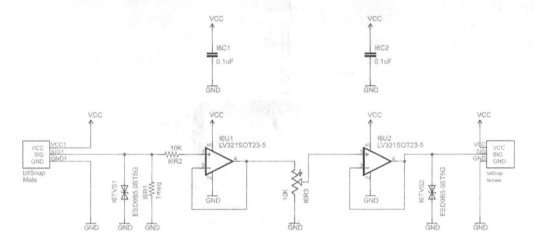

Figure 5.30 – Electronic interface circuits for the slider dimmer control

The electrical-electronics circuit schematic diagram for the M5Stack Core and the slider dimmer control is shown in *Figure 5.31*.

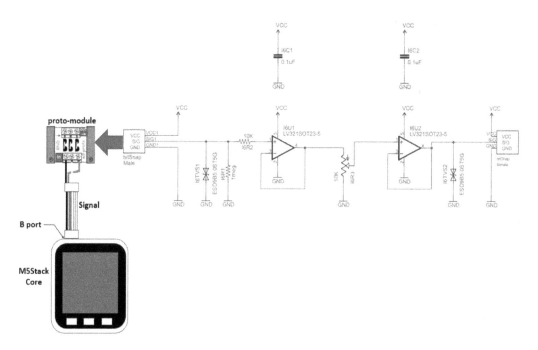

Figure 5.31 – Electronic circuit schematic diagram for the M5Stack Core and slide dimmer control

With the jumper harness wired to the proto module, you can attach the slide dimmer control to the assembled electrical interface unit. *Figure 5.32* shows the slide dimmer attached to the proto module. *Figure 5.32* clarifies the electronic circuit schematic diagram provided in *Figure 5.31* in terms of the two littleBits components' magnetic attachment.

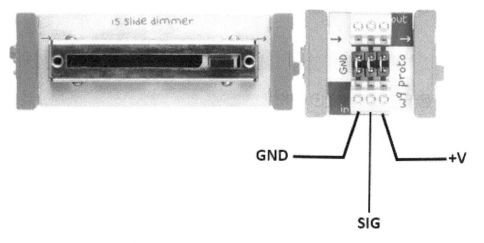

Figure 5.32 – Slide dimmer control for the proto module

This assembled electrical interface unit to the M5Stack Core completes the prototype build of the temperature sensor simulator device. *Figure 5.33* shows the complete temperature sensor simulator with a monitor. Congratulations – you have completed the assembly of your temperature sensor simulator with a monitor. The final step in this project is to program the device to display temperature values and provide a visual alarm upon exceeding a specified detection level.

> **Note**
> To provide electrical integrity and mechanical stability while adjusting the slide dimmer control, mount the operational amplifier-based potentiometer and the proto board on a **littleBits mounting plate**.

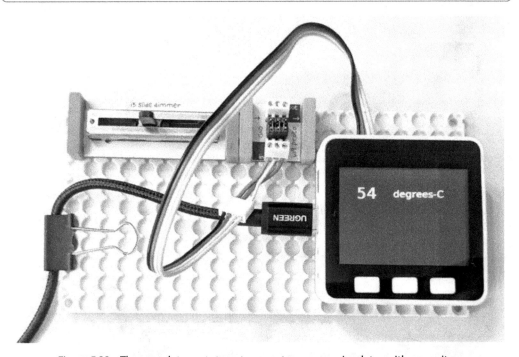

Figure 5.33 – The complete prototype temperature sensor simulator with a monitor

To create an aesthetically pleasing M5Stack Core UI, you will place the labels as shown in *Figure 5.34* onto the UiFlow layout. The UI will provide information on the temperature in °C, as well as displaying a visual alarm for exceeding a preset detection level. You can play with the **background color** and **font size** of the **UI labels** when executing the Blockly code.

> **Quiz 4**
> What are the full name description and the part number of the operational amplifier used on the littleBits slider dimmer control PCB?

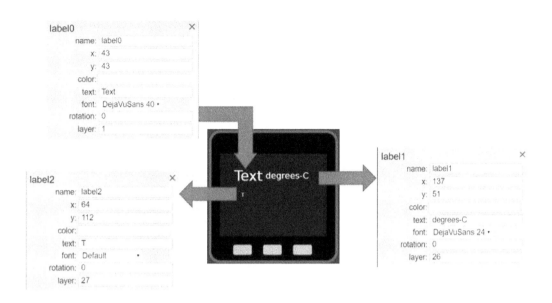

Figure 5.34 – Temperature sensor simulator monitor UI layout

You will use the Blockly code shown in *Figure 5.35* to activate your temperature sensor simulator monitor.

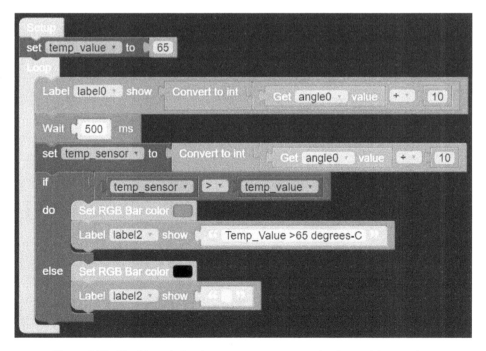

Figure 5.35 – Blockly code for the temperature sensor simulator with a monitor

By clicking the **run** button on the UiFlow taskbar and adjusting the slide dimmer control, you will see a range of temperature values displayed on the M5Stack Core's TFT LCD. *Figure 5.36* illustrates the operation of the temperature sensor simulator.

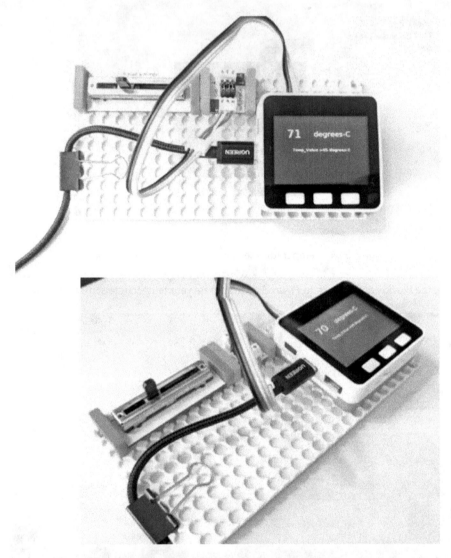

Figure 5.36 – Temperature sensor simulator with a monitor in operation

As you adjust the slide dimmer control, the LED bars will turn on upon exceeding the preset temperature value programmed within the Blockly code. Congratulations on building a temperature sensor simulator with a monitor! The final project you will build in this chapter is a littleBits LED flasher.

An M5Stack controller littleBits LED flasher

In this final project, you will learn how to wire a littleBits LED bar graph electronic module to the M5Stack Core. You will improve the interactive emoji UI discussed in *Chapter 3*. You will then obtain knowledge and skills in creating an aesthetically appealing visual animated display in this final project. Further, this final project will elucidate the key concept of using a block diagram with supporting electrical wiring diagrams and assembly pictures to assist you in building the LED flasher. The intention behind this learning and instructional approach is to allow you to assess the hands-on skills and technical knowledge you obtained from previous projects to build an M5Stack Core product. Therefore, you can think of this project as a capstone that allows you to assess and demonstrate your new knowledge and skills in building portable and interactive electronics using the M5Stack Core to your family, friends, and professional colleagues. *Figure 5.37* illustrates the littleBits LED flasher concept.

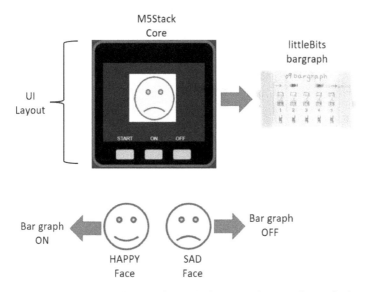

Figure 5.37 – Concept diagram for the littleBits LED bar graph LED flasher

You will notice that the UI layout has two illustrated faces – a happy and a sad one. You can turn on the LED bar graph by pressing the **ON** push button. The M5Stack Core's second pushbutton, or the **B** button, will turn on the LED bar graph. The happy face emoji will be displayed, along with the LED bar graph. The M5Stack Core's third pushbutton, or the **C** button, will turn off the LED bar graph. A sad face emoji will be displayed on the M5Stack Core's TFT LCD. You can start the flashing sequencing of the LED bar graph by pressing the M5Stack Core's first pushbutton, or the **A** button. The happy and sad face emojis will toggle, along with the flashing LED bar graph. You can find the happy and sad face images on GitHub here:

```
https://github.com/mrdon219/https-github.com-PacktPublishing-M5Stack-
Electronic-Blueprints
```

You will attach the LED bar graph using the littleBits bit snaps. The bit snaps allow the proto module and the LED bar graph to be magnetically and electrically attached to each other. The repulsive nature of the bit snaps lets you know when an attachment is incorrect – and a simple reversal will correct the assembly issue. *Figure 5.38* illustrates the proper orientation and assembly of the LED bar graph to the proto module. You will use the diagram to correctly attach the M5Stack electrical jumper harness to the proto module.

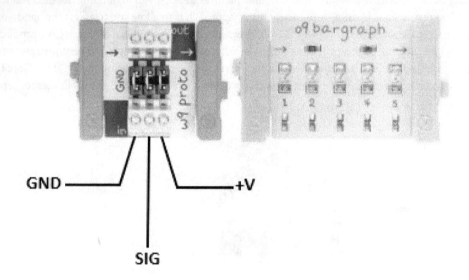

Figure 5.38 – Proper attachment and orientation of the littleBits LED bar graph to the proto module

Figure 5.39 provides further details on the electrical attachments of the jumper wire harness to the proto module to assist with correct assembly:

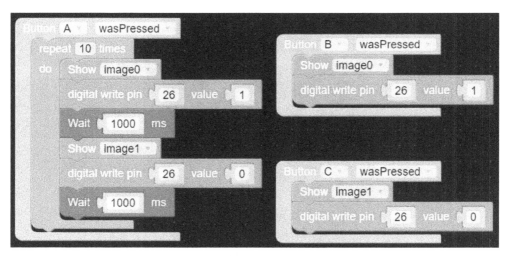

Figure 5.39 – Assembly of the jumper harness to the proto module

With the jumper harness attached and wired to the littleBits proto module, you can insert the **white connector** into the **B** port of the M5Stack Core. Place the littleBits electronic modules onto the plastic mounting plate to ensure proper electrical interfacing and mechanical rigidity are maintained during use. You can use the Blockly code to enable the LED flasher and the interactive emojis. Build the LED flasher Blockly code using *Figure 5.40* as a reference. Click on the **run** button on the taskbar with your mouse to engage with the LED flasher. *Figure 5.41* illustrates the M5Stack Core product in action.

Interactive quiz 5

Modify the Blockly code shown in *Figure 5.40* so that a counter is displayed to keep track of the number of LED flashes.

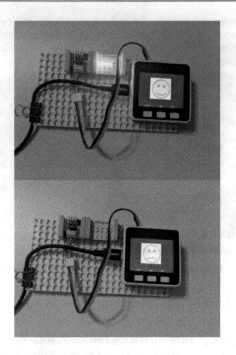

Figure 5.40 – The LED flasher Blockly code

Here are some pictures of the assembled and operational littleBits LED flasher.

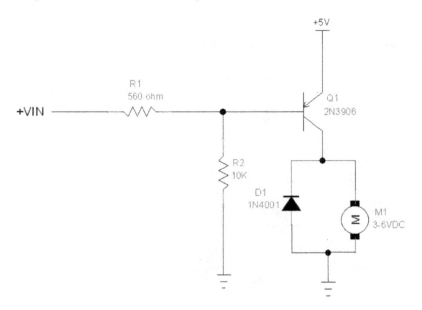

Figure 5.41 – The operational LED flasher device

Congratulations – you have successfully completed the M5Stack Core controller littleBits LED flasher project!

Summary

Congratulations – you have completed the hands-on activities and interactive quizzes in this chapter. In the chapter, you learned about the electronic circuits used to interface or connect discrete components to the M5Stack Core. You learned how to wire electronic circuits using solderless breadboards and proto modules. You learned how to build a virtual online circuit using Tinkercad Circuits to test a transistor LED driver circuit. You learned how to use a solderless breadboard to wire and test a transistor-operated DC motor. You learned how to create a physical simulator to behave like a temperature sensor using a littleBits slide control. When building a physical simulator, you learned about the NTC and its application with a thermistor.

Further, you learned how to use your coding skills and electronics knowledge to build a littleBits LED flasher. This project provided the concept diagram and the hardware and software components for building the M5Stack Core controller littleBits LED flasher. This project was different in its instructional approach, whereby you were assessed on the hands-on skills and technical knowledge obtained from previous chapters by building a functional and interactive product.

In this next chapter, you will take this knowledge of building an M5Stack Core controller to enable discrete solderless breadboard electronic circuits to work with an Arduino.

Interactive quiz answers

- Interactive quiz 1: Number 2
- Interactive quiz 2: Image
- Interactive quiz 3: Yes
- Interactive quiz 4: Yes
- Interactive quiz 5: Image

Quiz answers

Quiz 1: False

Quiz 2:

Total = I1 + I2

Total = (**7.61** + **5.47**) mA

Total = 13.08 mA (the correct value, so yes, they are correct)

Quiz 3: Text

Quiz 4: Single general-purpose, low-voltage, rail-to-rail output operational amplifier – LV321

6
M5Stack and Arduino

In the previous chapter, you learned the various electronic interfacing circuits that extend the **M5Stack Core** operational functions. You learned the various electronic interfacing circuits that allow **input and output** (**I/O**) control of electrical DC motors, discrete LEDs, and littleBits electronic modules. The littleBits electronic modules, such as a slide dimmer and bar graphs, are devices that can be operated with the M5Stack Core. You learned to program the M5Stack Core using the UiFlow Blockly code programming environment. Digital input circuits using a series resistor and a switch were discussed in the previous chapter. Discrete transistor switches using **negative-positive-negative** (**NPN**) and **positive-negative-positive** (**PNP**) configurations allow the operation of an electrical DC motor. You learned about these electronic interfacing circuits by reviewing the circuit schematic diagrams provided in *Chapter 5*.

In this chapter, you will learn how to enhance electronic interfacing techniques using an Arduino Uno. Further, the M5Stack Core2 is a touchscreen version of the original ESP32-based unit. With knowledge of creating a block diagram and electronic circuit schematic diagram obtained in *Chapter 3*, you will be able to wire the M5Stack Core2 to an Arduino Uno to operate various electronic circuit devices. You will be introduced to the M5Stack Core2 by coding various touchscreen UI controller applications and electronic modules by wiring them to the M5Stack Core2. Further, you will learn how to build an internet-based dashboard operated by an M5Stack Core2 and an Arduino Uno. An internet-based dashboard tool using flow data nodes will be presented in this chapter.

By the end of this chapter, you will be able to perform the following technical tasks with the M5Stack Core and solderless breadboarding:

- Code and test a basic M5Stack Core2 touchscreen UI controller
- Wire and test an Arduino-based digital inverting switch using a hex inverter **integrated circuit** (**IC**)
- Wire and test an Arduino-based digital counter device
- Wire and test an Arduino electromechanical relay controller

In this chapter, we are going to cover the following main topics:

- M5Stack Core2 touchscreen UI controls introduction
- Building a touch control inverting switch
- Building a touch control counter
- Building a touch control relay controller
- Building a touch control LED dimmer controller

Technical requirements

To engage with the chapter's learning content, you will need the **M5Stack Core2** to explore touchscreen controller applications and software coding of your UI designs. You will be introduced to an Arduino Uno **projects board** to allow ease in developing M5Stack Core2 touch control applications. Further, the UiFlow Blockly code software will be required to build and run the M5Stack Core2-Arduino Uno-compatible projects' touch controller applications.

You will require the following:

- M5Stack Core2
- Freenove Projects board
- One 4.7K ohm resistor
- One 74HC00 or 74LS00 Quad NAND Gate IC
- Three 10K ohm resistors
- One 220-ohm resistor

Here is the GitHub repository for software resources:

```
https://github.com/PacktPublishing/M5Stack-Electronic-Blueprints/
tree/main/Chapter06
```

M5Stack Core2 touchscreen UI controls introduction

The M5Stack Core2 is an enhanced version of the original Core unit. As with the original Core unit, the Core2 uses an **ESP32**-based microcontroller for providing internal hardware components such as the RGB bars, speaker, microphone, gyro, and accelerometers. An internal **vibration motor** is included with the M5Stack Coe2 unit. The M5Stack Core2 uses a specialized IC to operate a **capacitive touchscreen**. The touchscreen is overlayed with a **Thin Film Transistor** (**TFT**) LCD. *Figure 6.1* shows a conceptual diagram of the M5Stack Core2's touchscreen assembly. As shown in the conceptual diagram, the specialized IC is a touchscreen controller responsible for **touch detection** and processing of a human finger:

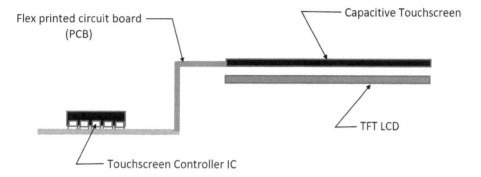

Figure 6.1: M5Stack Core2 touchscreen conceptual diagram

The touchscreen controller IC has a microcontroller that allows touch detection and processing actions. The flexible **printed circuit board** (**PCB**) allows touch detection and processing signals to be sent and received by the touchscreen controller IC. Further, the TFT LCD allows the user to interact with UI's visual imagery layout. The M5Stack Core2's UI layout is created using UiFlow Blockly code. The M5Stack Core2 is shown in *Figure 6.2*:

Figure 6.2: M5Stack Core2

The UI touchscreen buttons are located on the bottom portion of the M5Stack Core2 and are identified as three circles. Text can be placed above the three circles for personalizing specific control functions to be initiated by the M5Stack Core2. *Figure 6.3* illustrates the location of the M5Stack Core2 touchscreen buttons:

Touchscreen buttons

Figure 6.3: Touchscreen buttons

You can program the touchscreen buttons to provide various interactive control functions. Accomplishing this task requires accessing the UI palette within the UiFlow software. You will first need to establish communication with the M5Stack Core2. To do so, connect the M5Stack Core2 to your development machine using the **USB-C cable**. Insert one end of the USB-C cable into the M5Stack Core2. The other end of the USB-C cable you will insert into your development machine. The UiFlow **splash screen** will be visible on the M5Stack TFT screen. You will touch the **flow button** to advance to the **choose mode** screen. Touch the USB button, then the M5Stack Core2 will reset. After the unit resets, it will be in **USB mode**. *Figure 6.4* shows the steps for setting the M5Stack Core2 into USB mode:

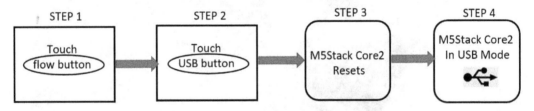

Figure 6.4: Setting M5Stack Core2 into USB mode

Quiz 1

_____ has a microcontroller that allows touch detection and processing actions to occur.

Once you have completed the steps, the M5Stack Core2 will have a visual image and message, as displayed in *Figure 6.5*:

Figure 6.5: M5Stack Core2 in USB mode

> **Note**
>
> Here are the M5Stack Core2 power management operations:
>
> **Power on**: One click with the power button on the left
>
> **Power off**: Long press the left power button for 6 seconds
>
> **Reset**: Click the RST button on the button side

Congratulations on setting up the **USB communications mode** for the M5Stack Core2! You will now proceed to provide a data connection with the UiFlow software. Open the UiFlow software and select the appropriate **common (COM)** port and the M5Stack Core2 device. Click the **OK** button after you have selected your COM port and M5Stack Core2 device. *Figure 6.6* shows the selection of the COM port and the M5Stack Core2 device:

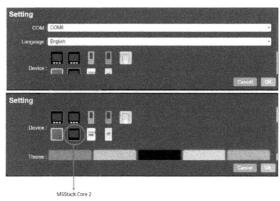

Figure 6.6: COM port and the M5Stack Core2 device selection

> **Note**
>
> You can use **Device Manager** to determine your COM port. Of the three visible devices under **Ports (COM & LPT)**, select the **Silicon Labs** COM port!

You can easily create a touch control application with M5Stack Core2 by placing a button on the UI layout. With the button placed, you will select from the UI Blockly code palette the **Button** option and the `touch_Button0_was pressed` code block. *Figure 6.7* shows the selection of these coding items:

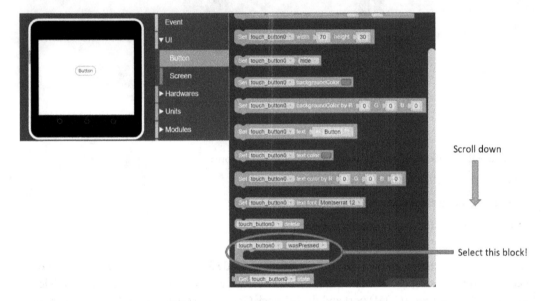

Figure 6.7: UI Button option and code block

Place the `touch_Button0_wasPress` code block onto the programming screen. You will now proceed to build a basic Hello World application. The Hello World application simply works by pressing the button, displaying the Hello World text, and turning on the red bars. *Figure 6.8* shows the Hello World code blocks. Click the **RUN** button to activate the application. Touch the button to display `Hello World!!` and turn on the red bar:

Figure 6.8: Touch-activated Hello World application

Congratulations—you have built your first interactive M5Stack Core2 touch control device! You will now add another button to turn off the red bar and clear the Hello World message from the TFT LCD. The UI for this new control feature is shown in *Figure 6.9*:

> **Interactive quiz 1**
>
> Using the Blockly code in *Figure 6.8*, what visual effect will be seen when the **Label-show** and the **Set RGB Bar Color** blocks are reversed?

Figure 6.9: Hello World – enhanced UI controls

To see the visual effect of the Hello World message display on the TFT LCD and the red bar turned on, click the **Run** button on the UiFlow taskbar. Touch the **Message** button to see the visual effects in action. Clicking the **No Message** button clears the TFT LCD and turns off the red bar. *Figure 6.10* illustrates the visual effects of the enhanced Hello World application:

Figure 6.10: The executed Hello World application

You now have the technical skills to proceed to the next project, whereby a touch control inverting switch will be constructed. The M5Stack Core2 will be wired to an Arduino Uno, equivalent to creating an inverting switch. Again, congratulations on completing the touchscreen UI controls introduction!

Building a touch control inverting switch

With an understanding of the M5Stack Core2's touchscreen setup and capabilities, you are now ready to build a touch control inverting switch. A touch control inverting switch is a digital device whereby a simple touch of a UI button will turn off an LED. Inverting switches aid **automation systems** whereby detecting an electrical contact open status of a normally closed electronic switch provides an inverted output state. Industry safety devices such as a pressure mat switch stop an **industrial process** when a machine operator stands on it. Upon standing on a pressure mat switch, its **normally closed (NC)** contacts are inverted or open, thus turning off the industrial process. With the industrial process stopped, a maintenance technician can work on an **automated machine** safely.

The touch control inverting switch consists of three primary components: the M5Stack Core2, an Arduino Uno compatible, a digital inverter IC, and an LED. The M5Stack Core2 will provide an input logic high-control signal to the Arduino Uno compatible mounted on the Freenove Projects kit. The output of the Arduino Uno compatible provides an equivalent logic high signal. The inverter IC will invert the input signal, providing an opposite output state. *Figure 6.11* illustrates the touch control inverting switch:

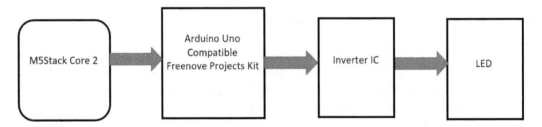

Figure 6.11: Touch control inverting switch

The M5Stack Core2 is capable of such output inversion using block code. The benefit of using this approach is to obtain technical knowledge in electronic circuit interfacing. Having electronic circuit interfacing skills allows the software developer or electronics maker to create hardware devices using this method of product creation. The Freenove Projects kit allows one to obtain electronic circuit interfacing skills by attaching jumper wires to single inline header pins mounted on the PCB. The Freenove Projects kit provides a variety of electronic mounted components such as an electromechanical relay, tactile pushbutton switches, a seven-segment LED display, an LED matrix, an I/O indicator, a 3V/5V DC power supply, and several potentiometers for hardware prototyping and experimentation. The Freenove Projects kit is shown next:

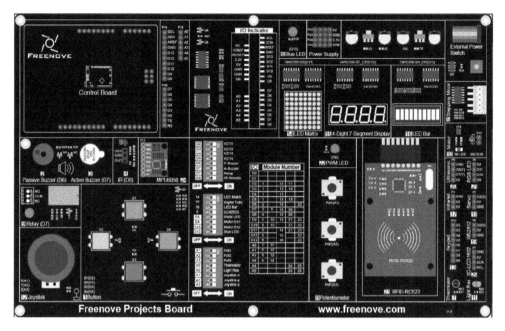

Figure 6.12: Touchscreen inverting switch

The Freenove Projects kit has an Arduino Uno compatible that serves as a **control board**. The Arduino Uno compatible has the same form or shape as a traditional Arduino board. Therefore, the original Arduino Uno board can be used with the Freenove Projects kit. To wire the M5Stack Core2 to the Freenove Projects kit to create a touch control inverting switch, an electrical wiring diagram is required. You will be able to create the inversion feature by including a digital inverter IC with the aid of an electrical wiring diagram. *Figure 6.13* shows a touch control inverting switch electrical wiring diagram:

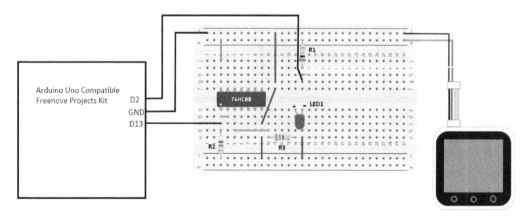

Figure 6.13: Touch control inverting switch

A 74HC00 or 74LS00 IC is a **quad NAND gate** performing the operation as an inverter by wiring two inputs of the discrete logic component. The 74HC00 or 74LS00 IC has 4 individual NAND gates in a 14-pin **dual-in-line (DIP)** package. One of the 74LS00 o r74HC00 NAND gates can be wired as an inverter by connecting two input pins of the logic IC. By applying an active high signal to the input of the inverter, the output level is logically low. An electronic circuit schematic diagram, as shown in *Figure 6.14*, provides an additional wiring reference diagram for the touchscreen inverting switch:

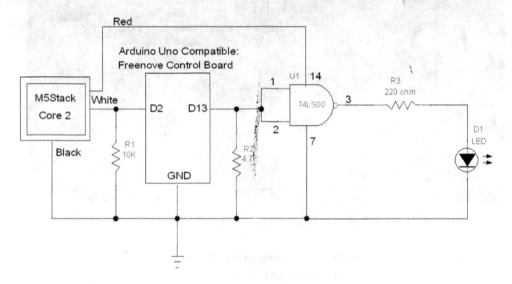

Figure 6.14: Touchscreen inverting switch electronic circuit schematic diagram

Further, the pinout and packaging of the 74LS00 or 74HC00 digital IC can aid you in wiring the touchscreen inverting switch, as shown in *Figure 6.15*:

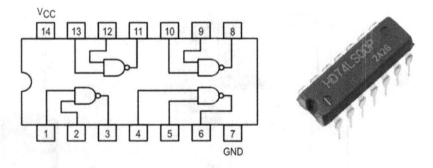

Figure 6.15: 74LS00 Quad NAND gate pinout and DIP digital IC

As shown in *Figure 6.16*, a physical **solderless breadboard** digital inverter IC circuit can be built. The electronic circuit schematic diagram illustrated in *Figure 6.14* will aid in wiring the breadboard circuit. The layout of the physical solderless breadboard components can be aligned with the electrical wiring diagram presented in *Figure 6.13*:

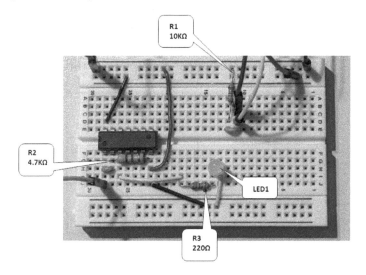

Figure 6.16: The completed solderless breadboard digital inverter IC circuit

The completed **prototype** circuit is shown in *Figure 6.17*:

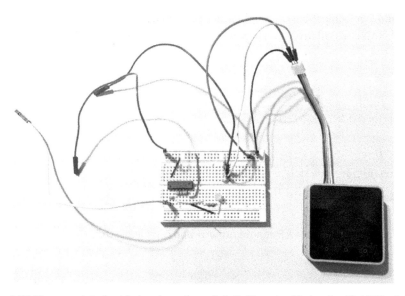

Figure 6.17: The completed solderless breadboard digital inverter IC circuit with M5Stack Core2

The next step in this project build is the electrical wiring of the solderless breadboard digital inverter IC circuit to the Freenove Projects kit. The electrical wiring consists of connecting three electrical wires from the solderless breadboard prototype to the Freenove Projects kit. The three electrical wires are the 74LS00/74HC00 input, ground, and the R1 10KΩ pulldown resistor. These three electrical wires connect to D13, ground, and D2 of the Freenove single inline header. *Figure 6.18* provides an electrical wiring table. In addition, you may use *Figure 6.13* as an additional electrical wiring reference diagram:

Solderless Breadboard Prototype	Arduino Uno Compatible Freenove Project kit
74LS00/74HC00 Input	D13
ground	ground
R1 10KΩ pulldown	D2

Figure 6.18: Electrical wiring table

Congratulations on wiring the M5Stack touch control inverting switch! You will now proceed to code the Freenove Arduino Uno compatible and M5Stack Core2. The Freenove Arduino Uno compatible uses C/C++ code to detect the M5Stack Core2-activated touchscreen buttons. Additionally, the Freenove Arduino Uno compatible will send the activated touchscreen control signal to the 74HC00 or 74LS00 quad NAND-based inverter. The C/C++ code is shown in *Figure 6.19*:

```
1   int M5StackPin = 2;  // The M5Stack Core 2 pin
2   int ControlPin = 13;     // The Inverter Output Control pin
3
4   void setup() {
5       pinMode(M5StackPin, INPUT);  // Set the M5Stack Core 2 pin as an input
6       pinMode(ControlPin, OUTPUT);    // Set the ControlPin as an output
7   }
8
9   void loop() {
10      if (digitalRead(M5StackPin) == HIGH) // if the M5Stack Core 2 UI ON button is pressed
11          digitalWrite(ControlPin, HIGH);         // Switch ON the Inverter Output ControlPin
12      else                                 // if the M5Stack Core 2 UI OFF button is pressed
13          digitalWrite(ControlPin, LOW);       // Switch OFF the Inverter Output ControlPin
14  }
```

Figure 6.19: Freenove Arduino Uno compatible code

The M5Stack Core2 block code is shown in *Figure 6.20*:

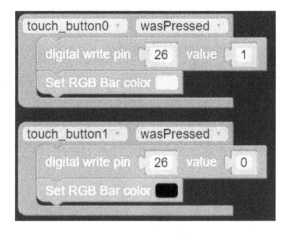

Figure 6.20: M5Stack Core2 block code

Finally, the UI for the M5Stack Core touch control inverter switch is shown in *Figure 6.21*:

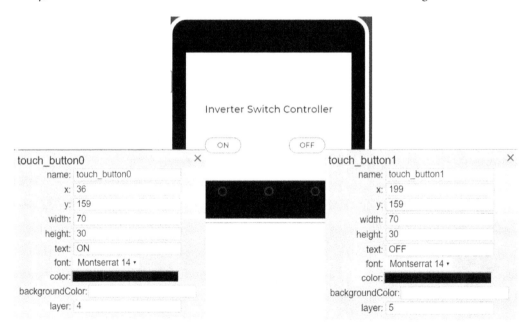

Figure 6.21: The M5Stack Core2 touch control inverter switch UI layout

Once the code has been installed on the M5Stack Core2 and Freenove Arduino Uno compatible units and the Projects kit board is powered on, the UI will be displayed on the touchscreen. *Figure 6.22* illustrates the active M5Stack Core2 UI touchscreen:

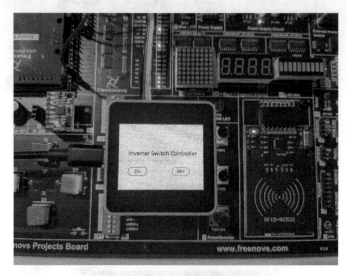

Figure 6.22: The M5Stack Core2 UI powered on

Touching the **OFF-UI** button will turn on the solderless breadboard LED. Touching the **ON-UI** button will turn off the solderless breadboard. This LED control operation is illustrated next:

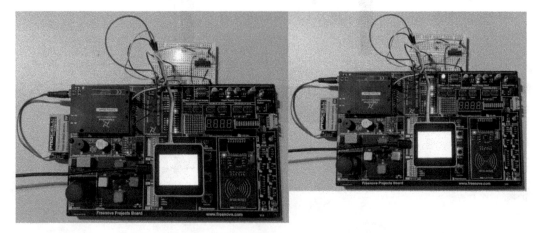

Figure 6.23: A functional M5Stack Core touch control inverter switch

You have completed the project and built a successful functional M5Stack Core touch control inverter switch. Great job! The circuit wiring skills obtained in this project will assist in building an M5Stack Core2 touch control counter.

Building a touch control counter

The touch control counter is a fascinating project where a seven-segment LED display feature will be built and demonstrated to the reader. There are four components required to make the touch control counter operational. The four touch control counter components are the M5Stack Core2, the Freenove Projects kit Arduino Uno compatible, a transistor relay driver circuit, and a **four-digit seven-segment LED display**. *Figure 6.24* shows a touch control counter block diagram:

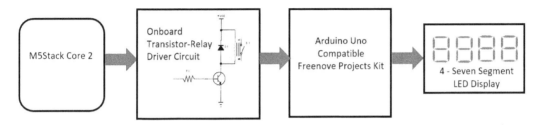

Figure 6.24: M5Stack touch control counter

The M5Stack Core2 provides a UI for operating the Freenove Projects kit onboard transistor relay driver circuit. The electromechanical relay **single pole single throw normally open (SPST NO)** contact provides a ground or low-control signal to the Freenove Projects kit Arduino Uno compatible. The Arduino Uno compatible will provide the latch, data, and clock signals to operate the four-digit seven-segment LED display circuit, thus allowing it to increment its displayed value. The displayed value increments every second. The first step in building the M5Stack touch control counter is to wire the ESP32-based controller to the Freenove Projects kit Arduino Uno compatible. *Figure 6.25* illustrates the electrical wiring interface between the M5Stack Core2 to the Freenove Projects kit Arduino Uno compatible:

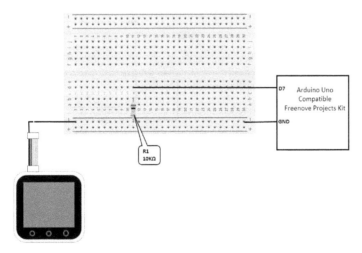

Figure 6.25: M5Stack Core2 to Freenove Projects kit electrical wiring diagram

The **digital pin D7** is the Freenove Projects kit's onboard transistor relay driver circuit. To enable the circuit, you will need to slide the DIP switch to the **ON** position. With the DIP switch in the **ON** position, touching the M5Stack Core2 **ON UI** button will operate the onboard transistor relay driver and its contacts:

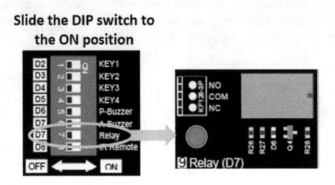

Figure 6.26: Enabling the transistor relay driver circuit

The final hardware build step is to wire the transistor relay driver circuit's *COM* contact to the ground. You will wire the transistor relay driver circuit's *NO* contact to the Freenove Projects kit Arduino Uno compatible digital pin 11. *Figure 6.27* illustrates this wiring scheme. As seen in *Figure 6.27*, a small solderless breadboard will be used for wiring the M5Stack Core2, the R1 10KΩ pulldown resistor, and the Arduino Uno-compatible components together:

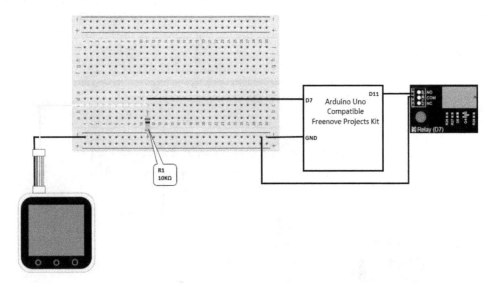

Figure 6.27: Wiring the transistor relay wiring contacts to the Freenove Projects kit

The final prototype build is shown in *Figure 6.28*:

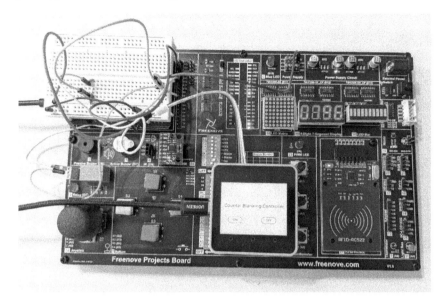

Figure 6.28: M5Stack Core2 touch control counter

Congratulations on building the touch control counter! As shown in *Figure 6.28*, the M5Stack Core2 UI has been modified. The UiFlow Blockly code used in the previous project will be implemented in this control device scheme. Refer to *Figure 6.20* for the UiFlow Blockly code. You will make a slight modification to the software name, which will change the label UI. The label UI will be named `Counter Blanking Controller`, as shown next. With the Blockly code modified, you may run it to display the new UI:

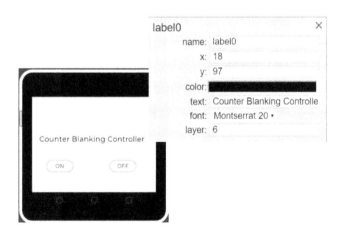

Figure 6.29: Modified label UI

The last step to completing the software phase is to upload the counter code to the Arduino Uno compatible. The counter code will manage the four-digit seven-segment LED display by incrementing the count value every 1 second. You may reset the counter to display four zeros using the **Reset** button on the Arduino Uno compatible. The code for the counter is shown in *Figure 6.30*.

With the counter code uploaded to the Arduino Uno compatible, you may test the complete system functionality. You will use the Arduino IDE to perform the Arduino Uno-compatible programming task of the project. As briefly mentioned in the introduction, the operation of the M5Stack Core2 controller is to blank the four digits of the seven-segment LED display. For a control-signaled optoelectronic component wired to a digital counter IC, there are three basic input pins to test or operate it. The three pins are the **blanking (BL)**, **lamp-test (LT)**, and **latch enable (LE)** pins. The LE pin, when pulled high, latches or freezes the LED display. The LT pin illuminates all seven segments of the optoelectronic component, thus testing it. The BL pin allows turning off the display. This feature is accomplished by pulling the BL pin low:

```
1    int latchPin = 12;        // Pin connected to ST_CP of 74HC595(Pin12)
2    int clockPin = 13;        // Pin connected to SH_CP of 74HC595(Pin11)
3    int dataPin = 11;         // Pin connected to DS of 74HC595(Pin14)
4    int countValue = 0;       // Digital tube display value
5
6    byte num[] = {0xc0, 0xf9, 0xa4, 0xb0, 0x99, 0x92, 0x82, 0xf8,//0-7
7                   0x80, 0x90, 0x88, 0x83, 0xc6, 0xa1, 0x86, 0x8e
8                  };//8-F
9
10   void setup() {
11     // set pins to output
12     pinMode(latchPin, OUTPUT);
13     pinMode(clockPin, OUTPUT);
14     pinMode(dataPin, OUTPUT);
15   }
16
17   void loop() {
18     for (int j = 0; j < 250; j++) {
19       ShowCount(countValue);
20     }
21     countValue++;
22   }
23
24   void ShowCount(int value) {
25     DigitalTube_MSBFIRST(0, num[value % 10000 / 1000]); delay(1);
26     DigitalTube_MSBFIRST(1, num[value % 1000 / 100]);   delay(1);
27     DigitalTube_MSBFIRST(2, num[value % 100 / 10]);     delay(1);
28     DigitalTube_MSBFIRST(3, num[value % 10]);           delay(1);
29   }
30
31   void DigitalTube_MSBFIRST(int number, byte value) {
32     digitalWrite(latchPin, LOW);
33     shiftOut(dataPin, clockPin, MSBFIRST, 0x01 << number);
34     shiftOut(dataPin, clockPin, MSBFIRST, value);
35     digitalWrite(latchPin, HIGH);
36   }
```

Figure 6.30: Four-digit counter code

The BL feature will be performed using the M5Stack Core2 counter blanking controller UI. Upon touching the **ON UI** button, the transistor relay driver circuit's *NO* closing contacts will provide a ground at digital pin D11. With digital pin D11 pulled low, the counter's display will turn off, thus going into blanking mode. *Figure 6.31* illustrates the blanking feature:

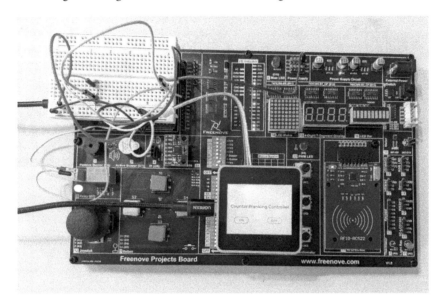

Figure 6.31: Implemented blanking feature

You will notice the blue LED indicator turn on when the transistor relay driver circuit is operating. This visual indicator provides immediate feedback that the blanking mode by way of the transistor relay driver circuit is working properly. Congratulations on completing the M5Stack Core2 touch control counter project! The transistor relay driver circuit will be explored further in the next M5Stack Core2 project.

Building a touch control relay controller

You now have the technical skills and knowledge from the previous project to build a touch control relay controller. The concept for the touch control relay controller is that the M5Stack Core2 will provide a nice UI to operate the Freenove Projects kit onboard transistor relay driver. The Freenove Projects kit onboard **transistor relay driver circuit** is built using five basic electronic/electromechanical components. The five basic electronic/electromechanical components are two fixed resistors, an LED, an NPN transistor, a silicon diode, and a **single-pole double-throw** (**SPDT**) electromechanical relay. An electronic circuit schematic diagram is illustrated in *Figure 6.32*:

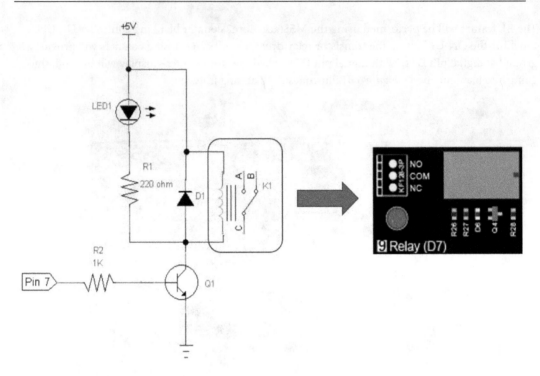

Figure 6.32: Onboard transistor relay driver circuit schematic

The convenience that the Freenove Projects kit provides is that the transistor relay driver is completely populated and assembled on the PCB. Such convenience allows M5Stack Core2 controller concepts to be easily prototyped and tested with the minimum wires. The ease with which interfacing the M5Stack Core2 controller with an Arduino Uno compatible makes developing unit control devices enjoyable and fun.

Upon digital pin 7 receiving a control signal from the Arduino Uno compatible, the NPN transistor Q1 will turn on. With the NPN transistor turned on, electrical current will flow from the +5V supply rail through the electromechanical relay's coil and the collector and emitter leads to the ground. The electromechanical relay coil will be energized, magnetically attracting the NO contacts to close. The silicon diode D1 prevents back **electromotive force (EMF)** produced by the stored electrical charge in the electromechanical relay's coil from flowing through the NPN transistor upon the de-energized circuit or being turned off. LED1 will turn on upon the transistor being biased from the pin 7 control signal. The resistor R2 limits the amount of biasing current flowing through the base, thus preventing burnout of the NPN transistor. The series resistor R1 limits the current flowing through the LED.

Note

Biasing is the DC operating or conducting point for a transistor.

To wire the touch control relay controller, an electrical wiring diagram is provided in *Figure 6.33*. As you observe, the electrical wiring diagram is a modification of the circuit shown in *Figure 6.27*. One obvious difference is that the onboard transistor relay driver circuit is operated by the Arduino Uno compatible. In *Figure 6.27*, the M5Stack Core2 operates the onboard transistor relay driver circuit instead:

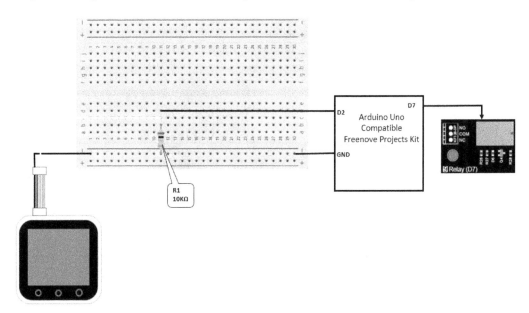

Figure 6.33: Touch control relay controller electrical wiring diagram

As a reference, the final prototype build is shown in *Figure 6.34*. Placement of the solderless breadboard onto the Arduino Uno compatible allows the creation of a convenient development platform for troubleshooting and product modifications. You will have noticed a similar approach to the placement of the M5Stack Core2 unit. With the Freenove Projects kit board, your M5Stack Core2-based Arduino Uno-compatible devices are easily transportable using this component placement scheme.

Quiz 2

In reviewing the electrical wiring diagram shown in *Figure 6.33*, what purpose does resistor R1 serve?

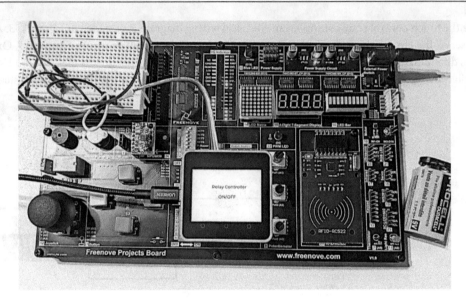

Figure 6.34: The complete touch control relay controller prototype

You will make a small modification to the touch control counter's UI by changing the label and event UIs. You will change the **UI** label to **Relay Controller**. The event UI will be changed to a slide switch. *Figure 6.35* shows the modified UI layout for the touch control relay controller. You will use a basic label UI for the **ON/OFF** text:

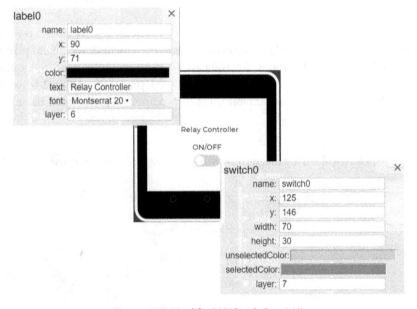

Figure 6.35: Modified M5Stack Core2 UI

The Blockly code to use to allow the slide switch to engage with the Arduino Uno compatible is shown next:

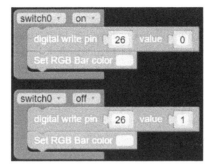

Figure 6.36: Modified M5Stack Core2 UI

The Blockly code will allow the operation of the onboard transistor driver relay by the UI slide switch positioning. This UI slide switch binary value is received by the Arduino Uno compatible digital pin 2. The C/C++ code that reads this digital pin binary value is shown in *Figure 6.37*:

```
1    int relayPin = 7;      // the number of the relay pin
2    int buttonPin = 2;     // the number of the push button pin
3
4    int buttonState = HIGH;      // Record button state, and initial the state to high level
5    int relayState = LOW;        // Record relay state, and initial the state to low level
6    int lastButtonState = HIGH;  // Record the button state of last detection
7    long lastChangeTime = 0;     // Record the time point for button state change
8
9    void setup() {
10       pinMode(buttonPin, INPUT);   // Set push button pin into input mode
11       pinMode(relayPin, OUTPUT);   // Set relay pin into output mode
12       digitalWrite(relayPin, relayState); // Set the initial state of relay into "off"
13       Serial.begin(9600);          // Initialize serial port,and set baud rate to 9600
14    }
15
16    void loop() {
17       int nowButtonState = digitalRead(buttonPin); // Read current state of button pin
18       // If button pin state has changed, record the time point
19       if (nowButtonState != lastButtonState) {
20          lastChangeTime = millis();
21       }
22       // If button state changes, and stays stable for a while, then it should have skipped the bounce area
23       if (millis() - lastChangeTime > 10) {
24          if (buttonState != nowButtonState) {  // Confirm button state has changed
25             buttonState = nowButtonState;
26             if (buttonState == LOW) {    // Low level indicates the button is pressed
27                relayState = !relayState;       // Reverse relay state
28                digitalWrite(relayPin, relayState); // Update relay state
29                Serial.println("Button is Pressed!");
30             }
31             else {                       // High level indicates the button is released
32                Serial.println("Button is Released!");
33             }
34          }
35       }
36       lastButtonState = nowButtonState; // Save the state of last button
37    }
```

Figure 6.37: The relay controller code

You will upload the relay controller code in *Figure 6.37* to the Arduino Uno compatible. The Blockly code shown in *Figure 6.36* will be executed on the M5Stack Core2 by clicking the UiFlow **Run** button. Ensure the DIP switch setting for enabling the Freenove Projects kit onboard transistor relay driver is on. Refer to *Figure 6.26* to enable the onboard transistor relay driver circuit. Touch the slide switch UI to turn on the relay. The blue LED will be on. *Figure 6.38* illustrates its control relay operation. Touch the slide switch UI and notice the blue LED turns off. Job well done; you have successfully built a touch control Relay controller. The final project of this chapter is the development of an LED dimmer controller. You will build an LED dimmer controller using the M5Stack Core2, the Freenove Projects kit Arduino Uno compatible, and an onboard discrete LED:

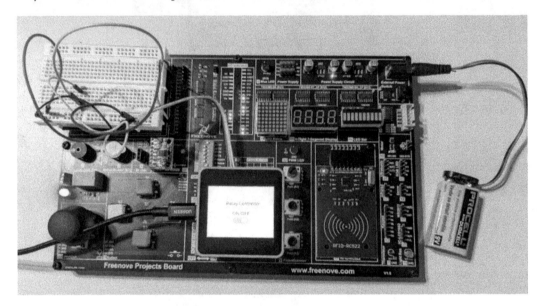

Figure 6.38: Onboard transistor relay driver LED

Building a touch control LED dimmer controller

This final project demonstrates controlling a discrete LED's light intensity with a varying output control voltage. The concept controller consists of three electronic components: the M5Stack Core2, the Freenove Projects kit Arduino Uno compatible, and a discrete LED circuit. The concept of an LED dimmer controller block diagram is shown in *Figure 6.39*:

> **Note**
> Applying an adjustable voltage source affects the electrical current flowing through an LED, thus changing the light intensity of the optoelectronic component.

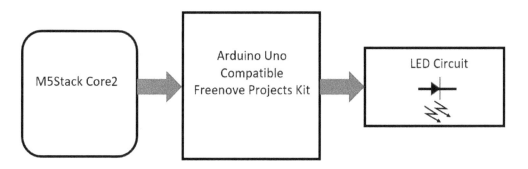

Figure 6.39: LED dimmer controller block diagram

You will use a **slider UI** to adjust the light intensity of the LED. The equivalent electronic component used to change a basic light source's light intensity is a **rheostat**. A rheostat is a variable resistor that can change the electrical current through adjusted resistances. Today's basic light source is the LED. The M5Stack Core2 and the slider UI will allow the LED light intensity to be adjusted by a slide to the right and left of the TFT LCD screen. You will replace the slide switch UI of the touch control relay controller with the slider UI. *Figure 6.40* illustrates the slider UI on the M5Stack Core2:

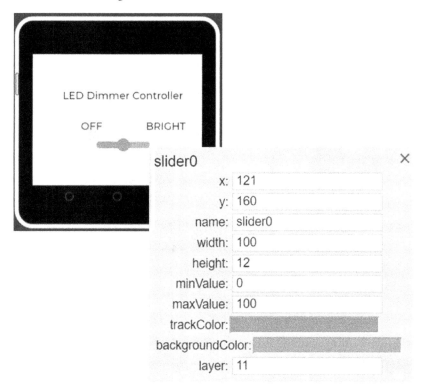

Figure 6.40: The slider UI included on the M5Stack Core2

There are two blue LEDs on the Freenove Projects kit. You will use the blue LED electrically wired to digital pin 13. To ensure the blue LED will be electrically wired to the Arduino Uno compatible, slide the DIP switch to the **ON** position. *Figure 6.41* shows enabling the blue LED wired to digital pin 13:

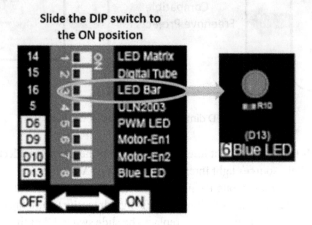

Figure 6.41: Enabling the digital pin 13 blue LED

With the blue LED circuit enabled and the M5Stack Core slider UI configured, you may build the dimmer controller software feature. The Blockly code is quite simple in construction yet effective in providing a varying output control signal. The control signal will provide varying discrete voltage values the Arduino Uno compatible's **analog channel 1 (A1)** will use to provide the appropriate LED intensity level. *Figure 6.42* shows the LED dimmer controller Blockly code:

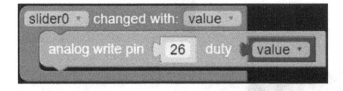

Figure 6.42: LED dimmer controller Blockly code

> **Note**
>
> The analog channel of the microcontroller converts varying signals to an equivalent binary value using an **analog-to-digital converter (ADC)**.

You can wire the M5Stack Core to the Freenove Projects kit Arduino Uno compatible using the electrical wiring diagram shown next:

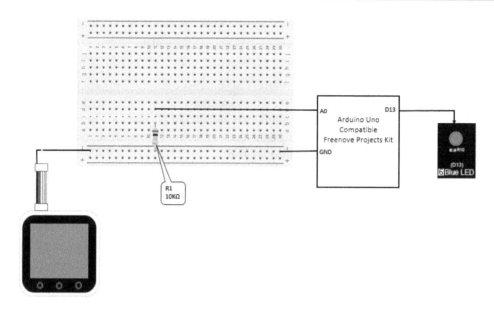

Figure 6.43: Touch control LED dimmer controller electrical wiring diagram

You will notice that the electrical wiring is like that shown in *Figure 6.33*. The convenient feature of using the Freenove Projects kit platform is the ease with which you can rapidly prototype new controllers using the Arduino Uno compatible as a smart interface controller. As shown in *Figure 6.43*, the solderless breadboard allows ease in changing electrical terminating points to the Arduino Uno compatible. This electrical wiring scheme allows new UI controller concepts to be rapidly developed and evaluated with ease. The final prototype build will appear as illustrated in *Figure 6.44*. With the M5Stack Core2 wired to the Freenove Projects kit Arduino, the Blockly code can be run on the ESP32-based controller:

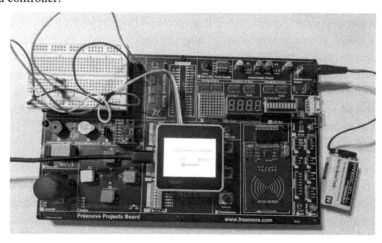

Figure 6.44: Touch control LED dimmer controller prototype

> **Quiz 2**
>
> Name the equivalent electronic component that the slider UI simulates.

You will complete the final step of the touch control LED dimmer controller by uploading the C/C++ code to the Arduino Uno compatible. The LED dimmer controller C/C++ code will allow the M5Stack Core2 analog controls signals to be read. The appropriate output control signals will be produced by the LED dimmer controller to provide the varying light intensity levels of the optoelectronic component. *Figure 6.45* shows the code for the LED dimmer controls:

```
1    int adcValue;        // Define a variable to save the ADC value
2    int ledPin = 13;     // Use D9 on Freenove UNO to control the LED
3
4    void setup() {
5      pinMode(ledPin, OUTPUT);              // Initialize the LED pin as an output
6    }
7
8    void loop() {
9      adcValue = analogRead(A1);            // Convert the analog of A0 port to digital
10     // Map analog to the 0-255 range, and works as PWM duty cycle of ledPin port
11     analogWrite(ledPin, map(adcValue, 0, 1023, 0, 255));
12   }
```

Figure 6.45: LED dimmer controller C/C++ code

With the code uploaded to the Arduino Uno compatible, adjust the LED light with the M5Stack Core2 slider UI. Sliding the UI control to the right increases the LED light intensity. *Figure 6.44* illustrates the function of the slider UI. Congratulations on successfully building a touch control LED dimmer controller!

Summary

Congratulations—you have completed the hands-on activities and quizzes in this chapter! In the chapter, you learned about electronic circuits interfacing to a Freenove Projects kit Arduino Uno compatible using the M5Stack Core2. You learn how to electrically wire basic electronic circuits using a solderless breadboard. You learned how to build touch controller applications using the M5Stack Core2. Touch controls such as buttons, slide switches, and the slider UI were explored by building various controller devices. The controller devices investigated in this chapter were an inverting switch, a counter, a relay controller, and an LED dimmer controller. The unique aspect of the inverting switch was the introduction of a quad NAND gate IC. You learned how to create an inverter circuit by electrically wiring two inputs of a NAND gate.

Further, you learned how to use your coding skills and electronics knowledge to build a blanking control for a four-seven-segment LED display. A transistor relay driver circuit controlled the data latch pin 11, thus blanking the four-digit seven-segment LED display. The chapter provided the concept diagram, hardware, and software components to build various M5Stack Core2 touch controllers. The projects differed in their instructional approach, whereby you assessed the hands-on skills and technical knowledge obtained from the previous chapters by building a functional and interactive product. In this next chapter, you will build upon the hardware circuit knowledge of building an M5Stack Core controller to build wireless devices using the M5Stack Core Bluetooth chipset.

Interactive quiz answers

Interactive quiz 1: The red LED bar will be displayed first, followed by the Hello World text.

Quiz answers

Quiz 1: A touchscreen controller

Quiz 2: It ensures the input pin of the M5Stack Core2 is not floating by providing 0V to it

Part 3: M5Stack IoT Projects

This part will focus on detecting Wi-Fi and Bluetooth signals and creating wireless controllers. This section intends to illustrate **Internet of Things (IoT)** concepts using the M5Stack Core controller. You will gain a hands-on understanding of IoT technologies by building and testing basic wireless control devices.

This part has the following chapters:

- *Chapter 7, Working with M5Stack and Bluetooth*
- *Chapter 8, Working with M5Stack and Wi-Fi*

7

Working with M5Stack and Bluetooth

You learned various approaches to wiring an M5Stack Core2 to an Arduino Uno compatible in *Chapter 6*. The **Freenove** Projects kit allowed the prototyping of an M5Stack Core2 to an Arduino quite easily. In the previous chapter, you learned how to create an inverting switch using a **digital quad NAND integrated circuit** (**IC**) with an Arduino Uno compatible. Further, a touch control counter, with various controllers, was built using the M5Stack Core2 as the UI for the Arduino Uno compatible. The Freenove Projects kit allows rapid development of these interactive devices. As observed in *Chapter 6*, the transistor relay driver provided an automation switch approach to interacting with the Arduino Uno compatible.

Besides using an **electromechanical relay** method of control, an ADC interfacing approach using the M5Stack Core2, in the LED dimmer controller project, M5Stack Core2 provided varying DC voltages used by the Arduino Uno compatible's **analog-to-digital converter** (**ADC**) channel. With such adjustable voltages, the M5Stack Core2 was able to change the Freenove Projects kit's LED light intensity. The slider UI was introduced with this project, thus allowing you the opportunity to explore a new approach to creating unique electronic controller product engagement activities using the M5Stack Core2.

In this chapter, IoT projects will be explored using the M5Stack Core2's Wi-Fi and Bluetooth chipsets. The intention of this chapter is to allow you to build unique wireless controllers using the ESP32 Bluetooth-embedded chipset. You will learn about **Bluetooth Low Energy** (**BLE**) technology by building wireless transmitters and receivers to operate RGB LEDs, sound-tone generators, and messenger devices. Further, you will learn how to work these wireless devices using a smartphone, tablet, or laptop computer.

By the end of this chapter, you will have learned how to perform the following technical tasks with M5Stack Core2 and BLE technology:

- Build an ESP32 BLE application

- Code and test an ESP32 BLE application

- Code and test a BLE transmitting device using the Nordic Semiconductor's nRF toolbox

- Send M5Stack messages to a BLE terminal app

- Create a Nordic Semiconductor nRF toolbox touchscreen UI and test an M5Stack Core2 BLE receiver controller

In this chapter, we are going to cover the following main topics:

- The ESP32 microcontroller Bluetooth chipset

- UiFlow BLE coding blocks pallet overview

- Detecting an M5Stack Core2 Bluetooth signal with the Nordic Semiconductor nRF toolbox

- Building an M5Stack Bluetooth messenger device

- Building an M5Stack Bluetooth receiver controller

- Creating a Bluetooth RGB LED light and sound generator

Technical requirements

To engage with the chapter's learning content, you will need the **M5Stack Core2** to explore touchscreen controller applications and software coding of your UI designs. You will be introduced to the Nordic Semiconductor nRF toolbox for BLE detection and control applications. Further, the UiFlow Blockly code software will be required to build and run the M5Stack Core2 BLE controller applications.

You will require the following:

- M5Stack Core2

- Nordic Semiconductor nRF toolbox

- Transistor relay module

- Solderless breadboard

- Tactile pushbutton switch

- Active piezo buzzer

- Jumper wires

- Standard connector-based jumper harness

- Smartphone
- Tablet

Here is the GitHub repository for software resources:

```
https://github.com/PacktPublishing/M5Stack-Electronic-Blueprints/
tree/main/Chapter07
```

The ESP32 microcontroller Bluetooth chipset

The M5Stack Core2 uses an ESP32-DOWDQ6-V3 microcontroller incorporating a 240 MHz, **dual-core** microprocessor. A dual-core microprocessor provides efficiency in computation and managing **input/output (I/O)** operations of the **microchip** using two **central processing units (CPUs)**. Within this family of ESP32 microcontrollers, the chip has a Bluetooth chipset consisting of a **link controller** and **baseband**. The Bluetooth link controller handles the **physical layer packets** and all communication timing. The link controller implements the link, the low-level **real-time** protocol that operates Bluetooth communications. The baseband is the physical layer of Bluetooth communications. The baseband manages physical channels and uses other services in communication such as security and **error correction**. The **baseband protocol** or operating rule performs the link controller within the Bluetooth chipset. *Figure 7.1* illustrates the Bluetooth chipset architecture:

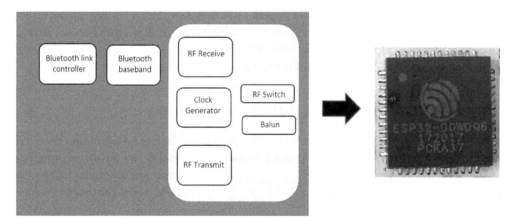

Figure 7.1: ESP32 Bluetooth chipset architecture

The **clock generator** within the Bluetooth chipset architecture is an **electronic oscillator** that produces a repetitive signal for synchronizing the Bluetooth link controller with the baseband. The **RF transmit circuit block** allows sending a **modulated signal** with the appropriate **carrier wave** and **intelligence data** to a designated or paired receiver. The **RF receive circuit** is responsible for obtaining intelligence data from a **demodulated** designated or paired transmitter signal. The **RF switch** is an electronic device used to route the 2.4 GHz signal received from a designated or paired transmitter. Finally, the **balun** is an electrical device that converts an unbalanced modulated received signal into a **balanced** or **differential** demodulated waveform. Traditionally, the balun is wired to the Bluetooth antenna to achieve a differential or balanced load for RF **signal integrity**. The balun electrical circuit concept is shown in *Figure 7.2*:

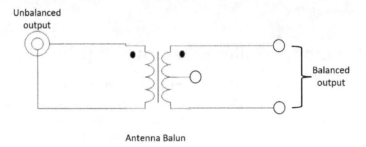

Figure 7.2: The antenna balun concept

Quiz 1

The clock generator is an electronic _____.

With an understanding of the ESP32 Bluetooth chipset, you will explore the UiFlow BLE coding blocks pallet. The BLE coding blocks pallet provides the operational means of allowing intelligence data to turn on or off an M5Stack RGB LED unit or transmit a text message. Therefore, the next section will provide an explanation of the BLE Blockly code blocks.

UiFlow BLE coding blocks pallet overview

The BLE coding blocks within the UiFlow software align with the **universal asynchronous receiver-transmitter (UART)** operation domain. A UART is a **microelectronics** device capable of providing **asynchronous serial communication** with a configurable data format and transmission speeds. The method of formatting the communication data is through a **parallel-serial conversion process** provided by the UART. The UART transmits the communication data serially to a receiving UART. Upon the receiving UART obtaining the communication data, it sends the digital information back to the transmitting UART in a parallel format. To accomplish this parallel-serial conversion process among receiving and transmitting UARTs, **transmit (Tx)** and **receive (Rx)** pins are used. *Figure 7.3* illustrates the physical wiring of the Tx and Rx pins of UART-based communication devices:

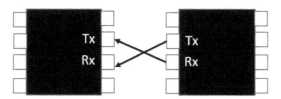

Figure 7.3: UART communications method

A final piece of information about the UART is data transmission without a digital clock. In **synchronous communications**, a clock allows for synchronizing digital system circuits sequencing simultaneously. In an asynchronous system, the digital system circuits are sequenced using start and stop bits. A data packet consisting of **intelligence** or **information** data with a start and stop bit allows the digital system circuits to operate in a uniform matter.

The UiFlow BLE Blockly coding blocks are centered on the UART for sending and receiving intelligence data packets. There are six primary BLE Blocky coding blocks within the pallet. *Figure 7.4* illustrates these:

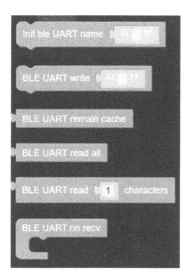

Figure 7.4: BLE-UART Blockly coding blocks

Here are descriptions for the BLE-UART Blockly coding blocks shown in *Figure 7.4*:

- `Init ble uart name`—Blockly code block used to initialize and configure wireless communication settings of the specified named Bluetooth device

- `BLE UART write`—Blockly code block used to send intelligence data using the BLE UART

- `BLE UART remain cache`—This Blockly code block is used to check the BLE UART number of data bytes

- BLE UART read—Reading the BLE UART cache data can be achieved with this Blockly code block

- BLE UART read characters—This Blockly code block allows reading the *n*-number of the BLE UART **cache** data

- BLE UART on recv—This Blockly code block allows decoding all BLE UART received data

> **Note**
>
> A cache (pronounced *cash*) is a hardware device or software element used to store data.

The primary UiFlow coding process of an M5Stack Core2 Bluetooth application using the BLE UART blocks consists of the following steps: initializing the BLE UART name, establishing the BLE UART conditional logic, and sending BLE UART data. To see the Blockly coding implementation of this process, you will begin by accessing the BLE UART pallet within the UiFlow software. The location of the pallet is shown in *Figure 7.5*:

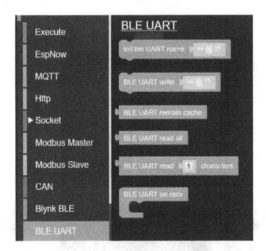

Figure 7.5: Accessing the BLE-UART Blockly code block pallet

You will then select BLE-UART Blockly code blocks from the pallet to build your specific **Internet of Things (IoT)** application within the UiFlow software. For example, select additional Blockly coding blocks to build the UiFlow BLE application shown in *Figure 7.6*:

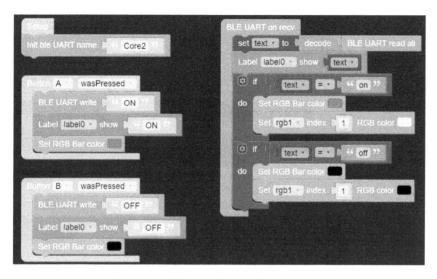

Figure 7.6: UiFlow BLE IoT application device

> **Note**
>
> The Green `rgb1` block requires the use of the RGB LED unit. You should add the RGB unit to see the coding blocks.

Notice that the BLE IoT application has alignment with the UiFlow coding process previously discussed.

> **Quiz 2**
>
> In reviewing *Figure 7.6*'s Blockly code blocks, upon pressing the **B** button, what information is sent by the UART?

Figure 7.7 shows the BLE IoT application alignment with the UiFlow coding process. As highlighted with callouts, you can rapidly develop a BLE IoT-based application using the UiFlow Blockly code software. In the next section, you will complete the build of the BLE IoT application device and test it using the Nordic Semiconductor nRF toolbox:

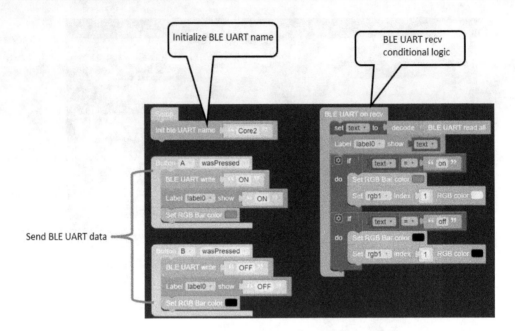

Figure 7.7: UiFlow coding process – BLE IoT application alignment

Congratulations on building the core UiFlow BLE IoT application!

Detecting an M5Stack Core Bluetooth signal with the Nordic Semiconductor nRF toolbox

The Blockly code application built aligns with the UiFlow BLE-UART development process discussed in the previous section. The next step in evolving this BLE IoT application is the use of a tool that will allow an investigation into sending wireless control commands and text messages on mobile devices such as a smartphone or a tablet. The Nordic Semiconductor nRF application is a toolbox that provides a wide range of popular BLE accessories and profiles.

Further, the Nordic Semiconductor nRF application profiles demonstrations of various profiles for health monitoring. Some nRF BLE profiles include **Heart Rate**, **Blood Pressure**, and **Proximity** monitors. These BLE monitors allow developers to explore UI/**user experience** (**UX**) designs and data aggregation methods. Such monitoring applications could be explored using the M5Stack Core2 as a **human-computer interaction** (**HCI**) development tool. The Nordic UART allows the development of text message and wireless control applications with the ESP32-based controller. *Figure 7.8* illustrates some of the UI-based BLE profiles of the Nordic Semiconductor nRF Toolbox app:

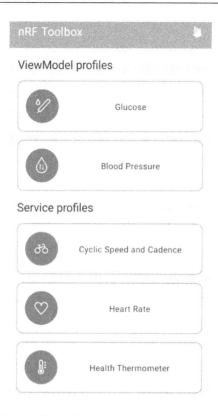

Figure 7.8: Example Nordic Semiconductor nRF Toolbox app profiles

The profile you will explore in this chapter will be the Nordic UART utility service app, shown next:

Figure 7.9: The Nordic UART utility service profile app

> **Note**
>
> The Nordic nRF Toolbox app is available for both Android and iOS mobile devices. Install the app and turn on Bluetooth prior to performing the following steps.

Before starting the Nordic UART utility service profile app, the M5Stack Core2 UI screen needs to be designed. Your UI screen can be designed as shown in *Figure 7.10*. You may customize the fonts and background screen to meet your visual needs. Further, the message may be changed to suit a more appealing ready prompt.

> **Quiz 3**
>
> Using *Figure 7.7*, the Nordic nRF Toolbox utility service app is associated with which UIFlow coding process?

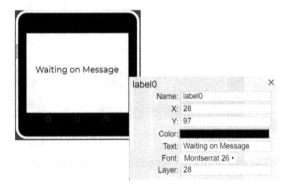

Figure 7.10: M5Stack Core2 UI layout

With the UI screen built, click the **RUN** button on the UiFLow Blockly code editor to execute the software shown in *Figure 7.7* on the M5Stack Core2 device. You should see the following text displayed on the M5Stack Core2 **Thin Film Transistor (TFT)** LCD:

Figure 7.11: M5Stack Core2 UI layout

You can open the UART utility service app from the installed Nordic nRF Toolbox software. Tap on the UART icon to see scanned BLE devices in your environment. Select the Core2-connected device from the list. *Figure 7.12* illustrates the scanned connected BLE devices:

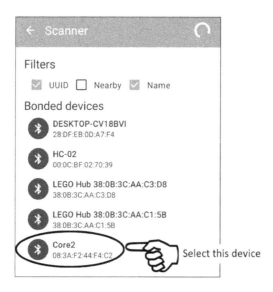

Figure 7.12: List of connected BLE devices

> **Note**
>
> The display may vary between an Android and iOS device.

Touching the Core2 listed device shown in *Figure 7.12* will display an output window, shown next:

> **Note**
>
> A **universally unique identifier** or **UUID** is a 128-bit label used for information in computer systems. Apple Computer in the 1980s originally used UUIDs in its **Network Computing System (NCS)**.

Figure 7.13: Nordic nRF Toolbox UART output window

Note

For iPhone users, select **Log** to see the Core2 device for selection and Bluetooth pairing with the nRF Toolbox app.

You will type in the text to send from the Nordic nRF Toolbox app to the M5Stack Core2 device, as shown in *Figure 7.14*. Touch the **Send** button on the app's UI to deliver a text message to the M5Stack Core2:

Figure 7.14: Sending Hello World text message to the M5Stack Core2 device

After pressing the **Send** button, a Hello World text message will be displayed on the output window and the M5Stack Core2 device. *Figure 7.15* illustrates a Hello World text message displayed on a smartphone and an M5Stack Core2 device:

Interactive quiz 1

Can low-level emojis such as :) be displayed on an M5Stack Core2 device?

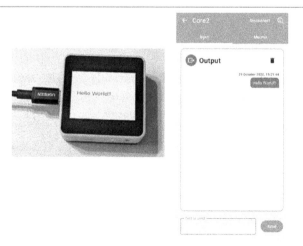

Figure 7.15: Hello World text message displayed on a smartphone and an M5Stack Core2 device

Congratulations on having the Nordic Semiconductor nRF Toolbox running successfully on your smartphone! You can install the Toolbox app on a tablet to achieve the same results. In the next topic of this chapter, you will explore the Blockly code blocks shown in *Figure 7.7* by enhancing the Bluetooth messenger's operation.

Building an M5Stack Bluetooth messenger device

You now have a working knowledge of using the Nordic Semiconductor nRF Toolbox app by setting up the app on a mobile device such as a smartphone or tablet. With the toolbox running effectively on your selected mobile device, you were able to detect and pair with the toolbox. The detection and pairing process was quite seamless, and as shown in *Figure 7.13*, you were able to send a text message to the M5Stack Core2. *Figure 7.16* provides a use-case diagram of the M5Stack Core2 as a basic Bluetooth messenger device:

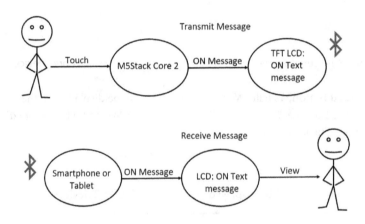

Figure 7.16: M5Stack Core2 use case – a basic messenger device

As observed in *Figure 7.16*, the user or actor initiates the messaging by entering or typing an ON message into the UART utility service profile of the Toolbox app. The message is sent or transmitted to the other actor by the M5Stack Core2 to the smartphone or tablet receiving the text. This send-and-receive process is the primary operation of the UART, as discussed in the *UiFlow BLE coding blocks pallet overview* section of this chapter.

In reviewing the Blockly code blocks in *Figure 7.7*, the transmit and receive approach of text messaging is quite simple. The BLE UART on recv loop's first Blockly code block decodes all Bluetooth messages sent by the M5Stack Core2 in a text format. The location of the received text message is captured by the label0 UI. The restriction of the text messages received by the smartphone or tablet is based on the `text` variable being specific to the on or off hardcoded string messages within the Blockly code blocks' programming structure. *Figure 7.17* illustrates the `BLE UART on recv` Blockly code blocks' programming structure. The primary structure of the Blockly code blocks is based on two conditional statements that act as message **aggregators** to detect specific or hardcoded text sent by the M5Stack Core2 device. With such a programming structure, any text message can be received by a smartphone or tablet quite easily:

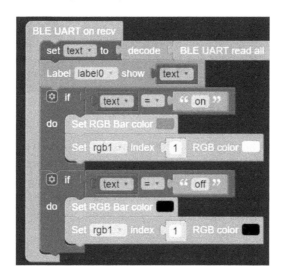

Figure 7.17: BLE UART on recv programming structure

> **Quiz 4**
>
> Identify the Blockly code blocks in *Figure 7.17* used to detect the hardcoded text message sent by the smartphone to the M5Stack Core2.

The Blockly code blocks responsible for sending text messages to a smartphone or tablet are shown in *Figure 7.18*. By touching the first circle (**Button A**) on the M5Stack Core2 TFT LCD, the text message `Mr. Tony Stark` will be transmitted to the smartphone or tablet. Touching the second circle (**Button B**), the text message `Mr. Tony Stark is Iron Man` will be wirelessly sent to the smartphone or tablet. `Figure 7.19` illustrates the event-based outputs associated with M5Stack Core2 **Button A** and **Button B**:

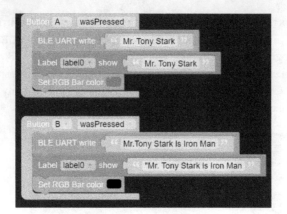

Figure 7.18: Transmitting Blockly code blocks

The BLE UART write Blockly code block allows a string or hardcoded text to be transmitted to a smartphone or tablet. Further, the RGB bar LEDs provide visual indicators of when the message has been transmitted. Although the selected colors used in this text messaging event have no significant meaning, for crucial or real-time operations, the appropriate color would be selected. The selected color would align with the critical messaging of the text—for example, red would align with an emergency status message such as help or stop:

Figure 7.19: BLE UART write transmitted string messages

As observed on your smartphone or tablet, the text messages have been sent and displayed on the Nordic Semiconductor nRF Toolbox app's output window. Another visual effect occurred, and that is the RGB bar is active during the text message transmitting events. Upon sending the Mr.Tony Stark text message, the red LED of the RGB bar is illuminated. Sending the Mr.Tony Stark is Iron Man text message turns off the RGB bar. These LED illumination effects are shown in *Figure 7.19*. Therefore, you can send a variety of text messages by changing the **string** contained within the BLE UART write Blockly code block. Congratulations on successfully building your M5Stack Core2 messenger device! By completing this project, you now have the knowledge and skills to send text messages from the M5Stack Core2 to a smartphone or tablet. With these technical competencies, you will be able to build an M5Stack Bluetooth receiver controller.

Building an M5Stack Bluetooth receiver controller

The M5Stack Core2 Bluetooth receiver controller is an enhancement of the messenger device you build in the previous project activity. The concept of sending a text message using the M5Stack Core2 to a smartphone or tablet was investigated. You will take the prototype built with the use-case diagram shown in *Figure 7.16* and convert it to receiver-controller concept diagrams shown in *Figures 7.20* and *7.21*. *Figure 7.20* illustrates an ON-control message controller:

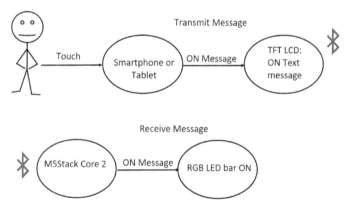

Figure 7.20: Sending an ON-control message

An OFF-control message controller use-case diagram is shown in *Figure 7.21*:

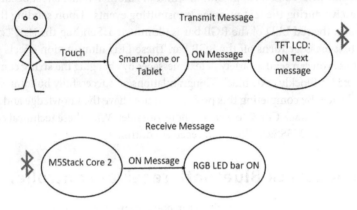

Figure 7.21: Sending an OFF-control message

Upon the user sending the control message (ON or OFF), the smartphone will display it on its touchscreen LCD. The M5Stack Core2 will display the message and turn the RGB LED bar on or off accordingly. The buttons on the M5Stack Core2 will be used to send ON- or OFF-control messages. *Figure 7.22* illustrates the Blockly code blocks to accomplish the wireless control operation:

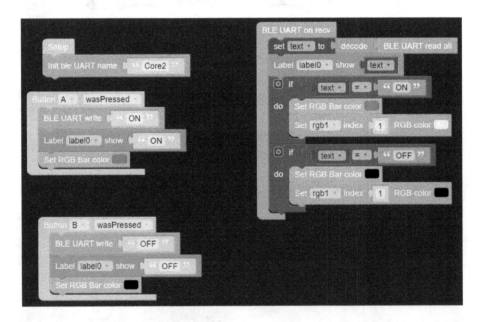

Figure 7.22: BLE-UART receiver-controller Blockly code blocks

Attach the M5Stack Core2 device to your development system using a USB-C cable. Select the correct COM port using the UiFlow software **Setup** feature. Click the **RUN** button with your mouse located on the bottom right of the UiFlow Blockly code block editor. Typing ON in the Nordic Semiconductor nRF toolbox text-to-send box and tapping the **Send** button will display the message on the smartphone or tablet. An ON text message will be displayed on the M5Stack Core2 device, with the red LED bar visible on each side of the BLE UART-based controller. *Figure 7.23* illustrates this text-received red LED bar **ON** operation:

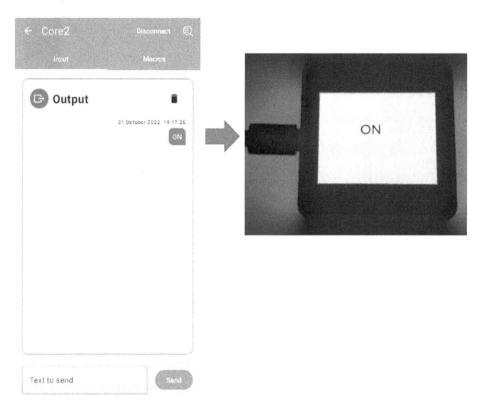

Figure 7.23: Text received red LED bar ON-control operation

Typing OFF in the Nordic Semiconductor nRF toolbox text-to-send box and tapping the **Send** button will display a message on a smartphone or tablet. An **OFF** text message will be displayed on the M5Stack Core2 device, with the red LED bar turned on each side of the BLE UART-based controller. *Figure 7.24* illustrates this text-received red LED bar **OFF** operation:

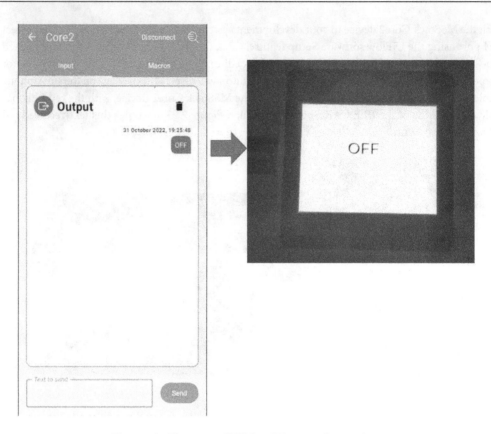

Figure 7.24: The text red LED bar OFF-control operation

If you were able to achieve the outcome results illustrated in *Figures 7.23* and *7.24*, congratulations on successfully building an M5Stack Bluetooth receiver controller! If your controller did not work successfully, review your Blockly code blocks for errors. With the errors corrected, repeat the steps of the wireless control shown in *Figures 7.23* and *7.24* to obtain the correct operation of your Bluetooth receiver controller.

You can enhance the Bluetooth receiver controller by adding a transistor relay module for **high-current** control operations. The Bluetooth receiver controller can be augmented to drive high-current **electromechanical** loads such as DC motors and solenoids. The switching **contacts** of the transistor relay module will be wired to these electromechanical loads operated by the M5Stack Core2 device.

Figure 7.25 shows a wiring diagram to accomplish this **augmented control**. The electrical wiring between the M5Stack Coe2 and the transistor relay module will require three individual jumper wires to make the **electrical interface** work properly. You will take three individual jumper wires and place them in line with the standard connector-based wire harness. These individual jumper wires will ensure the control signal voltage produced by the M5Stack Core2 device is present at the transistor relay module to turn it on:

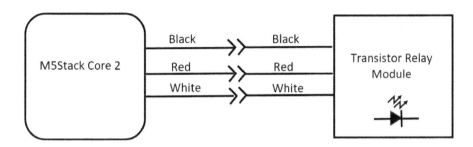

Figure 7.25: M5Stack Core2 to transistor relay module electrical wiring diagram

Figure 7.26 shows the inline jumpers connected to the standard connector-based wire harness:

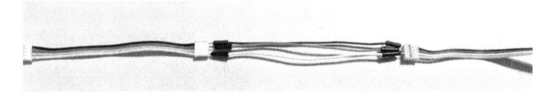

Figure 7.26: The inline jumper wires

You will modify the Bluetooth receiver-controller code to operate the transistor relay module with the M5Stack Core2 device. *Figure 7.27* illustrates the modified Blockly code to accomplish the transistor relay control operation. The minor change to the code is the use of the `digital write pin` blocks. The `digital write pin` block will turn ON the transistor relay module by writing a **binary 1** value to pin 26 of the standard connector-based wire harness. The **binary 1** value is equivalent to a DC voltage value of 3.3V. The `digital write pin` block will turn OFF the transistor relay module by writing a **binary 0** value to pin 26 of the standard connector-based wire harness. The binary 0 value is equivalent to a DC voltage value of 0V:

Interactive quiz 2

What effect does typing ON- and OFF-control commands in lowercase letters have on the control operation of the Bluetooth receiver controller?

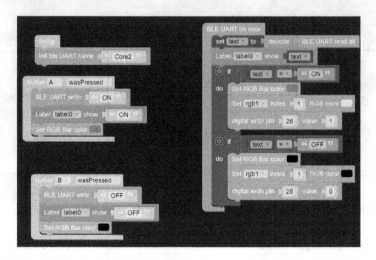

Figure 7.27: The digital write pin Blockly code blocks augment the Bluetooth receiver controller

Attach the modified jumper harness between the M5Stack Core2 device and the transistor relay module. Run the new code on the M5Stack Core2 device within the UiFlow Blockly code editor. Repeat the illustrated step shown in *Figure 7.25*. The M5Stack Core2 will display ON with the transistor relay module illuminating the red LED. *Figure 7.28* illustrates this Bluetooth receiver-controller operation step:

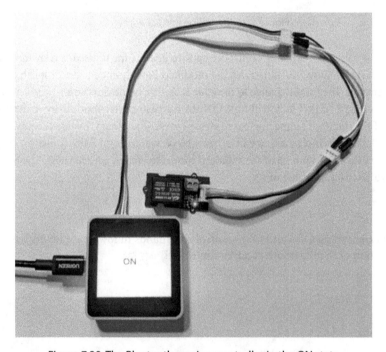

Figure 7.28: The Bluetooth receiver controller in the ON state

> **Quiz 4**
>
> In reviewing the Blockly code shown in *Figure 7.27*, identify the mystery unit.

Repeating the step illustrated in *Figure 7.25* will turn the transistor relay module and the onboard LED off. The M5Stack Core2 device will display an **OFF** text message during this smartphone or tablet interaction. This operation of the augmented Bluetooth receiver-controller step is illustrated in *Figure 7.29*:

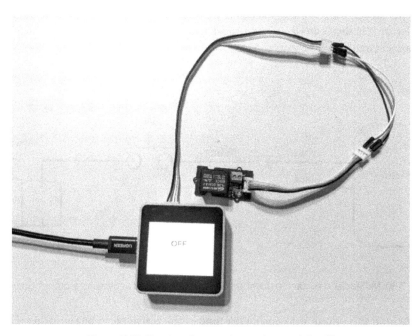

Figure 7.29: Turning the transistor relay module off

Congratulations on completing the augmented Bluetooth receiver-controller project! You now have the knowledge and hands-on skills to create a wireless controller capable of operating high-current electromechanical such as a DC motor or solenoid. The final project in this chapter is an offshoot of the previous two Bluetooth receiver controllers. This final project will use RGB LED bars with an active piezo buzzer to produce sound. Further, there will be an exploration of creating an electromechanical Morse code key in the last section of this chapter.

> **Interactive quiz 3**
>
> Try using a different control-command text message within the Nordic Semiconductor nRF toolbox, such as a 1 and 0. What effect do these new control-command text messages have on the Bluetooth receiver controller?

Creating a Bluetooth RGB LED light with sound

In the final project of this chapter, the Blockly code used in the previous activity will be used to operate an active piezo buzzer. When you send an ON text message using the Nordic Semiconductor nRF toolbox, the M5Stack Core will turn on the red LED bar, along with blaring an active piezo buzzer. Conversely, sending an OFF text message will turn off the red LED bar and the piezo buzzer. The jumper wire harness provided in *Figure 7.26* will be modified by using two wires instead of three. The second standard connector-based jumper is not required for this project. One unique change to the digital output pin 26 is the attachment of a tactile pushbutton switch wired in series with the active piezo buzzer. *Figure 7.30* shows an electrical wiring diagram of the tactile pushbutton switch-activated active piezo buzzer circuit:

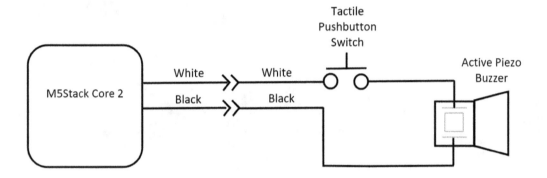

Figure 7.30: M5Stack Core2 tactile pushbutton switch-activated active piezo buzzer circuit

The concept behind this circuit is to enable the tone of the active piezo buzzer using the Nordic Semiconductor nRF Toolbox app. With the circuit turned ON by the Toolbox app, the tactile pushbutton switch will allow the tone to be heard with a single press. The user, therefore, can turn the ON or OFF tone with a single press of the tactile pushbutton switch. The red RGB LED is not affected by the tactile pushbutton switch due to the **optoelectronic emitter** being internally hardwired. The final project build prototype is illustrated in *Figure 7.31*.

> **Note**
> An active piezo buzzer uses a common DC voltage to operate the sound-producing electrical component.

The easter egg product with this project is a Morse code oscillator. You can create the *dit-dah* sounds of a Morse code oscillator by the length of pressing the tactile pushbutton switch. The length of the momentary presses on the tactile pushbutton switch corresponds to the messaging of Morse code.

Therefore, with some practice, you will be able to send messages using Morse code. Once you have completed the practice, you can turn it off using the Bluetooth receiver-controller function.

A similar approach may be used with the transistor relay module. The transistor relay contacts will produce a mechanical *click* sound, thus emulating an old-time Morse code key. The electrical wiring diagram illustrated in *Figure 7.25* can be modified by placing the tactile pushbutton switch in line with the white wire. Enabling the electromechanical Morse code device can be accomplished by sending an ON text message with your smartphone or tablet:

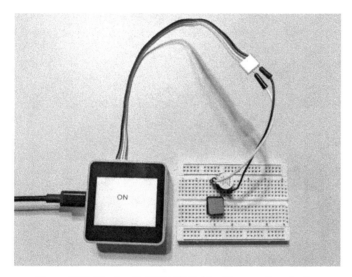

Figure 7.31: Bluetooth RGB LED light with sound controller

Figure 7.32 provides an electrical wiring diagram for the mechanical Morse code key:

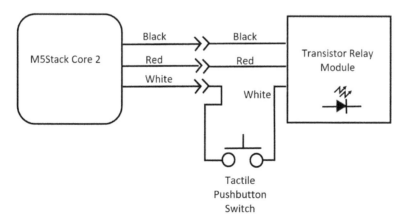

Figure 7.32: Electromechanical Morse code key wiring diagram

The new solderless breadboard prototype for the electromechanical Morse code key is shown in *Figure 7.33*:

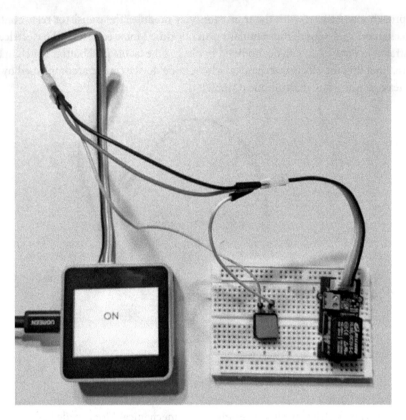

Figure 7.33: Electromechanical Morse code key solderless breadboard prototype

Final congratulations are in order! You have successfully completed creating a Bluetooth RGB light with sound project activities. You now have the technical knowledge and hands-on skills in building Bluetooth devices using an M5Stack Core2 device. Further activities and ideas to consider are to revisit some of the Snap circuits and Arduino Uno projects presented in previous chapters to add a Bluetooth wireless control feature. Expound on the interactive quizzes presented in the chapter to further explore Bluetooth wireless technologies.

Summary

Congratulations—you have completed the hands-on activities and quizzes in this chapter! In the chapter, you learned about the Bluetooth chipset integrated within the ESP32 microcontroller. You learned about the link controller and the baseband subcircuits that aid in the transmission of wireless data. The UiFlow Bluetooth coding blocks pallet was presented in the chapter. You obtained technical knowledge about the operation of a UART through a series of hands-on coding activities implemented on an M5Stack Core2 device. The UIFlow coding process was introduced in this chapter. With the UiFlow coding process, you learned about the BLE IoT application alignment of the key coding steps. The key coding steps learned in this chapter were `Initialize BLE UART Name`, `Send BLE UART Data`, and `BLE UART recv` conditional logic. BLE IoT application Blockly code blocks were introduced and implemented within this chapter. With the BLE IoT application, you were able to use the Nordic Semiconductor nRF toolbox to detect the UUID of the M5Stack Core2 device.

Further, you built a Bluetooth Messenger device using the BLE IoT application blockly code blocks and the Nordic Semiconductor nRF toolbox. With these wireless tools, you were able to send a text message to your smartphone or tablet and the M5Stack Core2 device. You learned how to convert the Bluetooth messenger device into a receiver controller. With the receiver controller, you were able to operate the M5Stack Core2 device RGB LED bar. You augmented the output operation of the controller by wiring a transistor relay module.

Finally, the Bluetooth receiver controller was electrically modified to operate an active piezo buzzer. To provide a silencing feature to an active piezo controller, you wired a tactile pushbutton switch to the control signal line of the M5Stack Core2 device. With the inline tactile pushbutton switch, you were able to create a practice Morse code tone oscillator. You then transformed the practice Morse code tone oscillator into an electromechanical key by wiring a transistor relay module. Pressing the tactile pushbutton switch to turn on and off the transistor relay module produced the mechanical *dit-dah* sound of a Morse code key.

In the final chapter of *M5Stack Electronic Blueprints*, you will become familiar with the Wi-Fi features included with M5Stack Core and Core2 devices. You will conduct Wi-Fi experiments to scan and detect wireless nodes. You will explore visual detection indicators and audible alarms in the chapter. You will be introduced to using the Arduino IDE platform to code C/C++ Wi-Fi-enabled detection devices.

Interactive quiz answers

Interactive quiz 1: Yes, based on emojis created using text characters.

Interactive quiz 2: The RGB LED bar will not turn on.

Interactive quiz 3: The Bluetooth receiver controller will not respond. Blockly code will need to be modified to accept the new text commands.

Quiz answers

Quiz 1: Oscillator

Quiz 2: OFF

Quiz 3: `Send BLE UART Data` and `BLE UART recv` conditional logic

Quiz 4: RGB unit

Working with the M5Stack and Wi-Fi

You learned about various approaches to wiring an M5Stack Core 2 to create different Bluetooth devices in *Chapter 7*. The ESP32 Microcontroller Bluetooth chipset was discussed in detail using a **system architecture block diagram**. A Bluetooth messaging device was built to explore the capabilities of the UiFlow BLE UART coding block pallets. Understanding the versatility of the coding blocks in allowing wireless control with the ESP32 Bluetooth chipset will help you use Wi-Fi technology in this chapter on Wi-Fi.

In this chapter, you will obtain hands-on skills and knowledge in Wi-Fi technology as it aligns with the IoT. You will explore hands-on approaches to using the M5Stack Core as a Wi-Fi receiver and controller to see networked wireless devices and operate them with a smartphone. You will learn how to create a **wireless access point** (**WAP**) with the M5Stack Core to control a Snap Circuits Bicolor LED. Furthermore, you will be introduced to the M5Stack Core **application programming interface** (**API**) by reviewing the C++ syntax. You will learn more about the M5Stack Core API by enhancing the existing C++ code.

The electronic circuit interfacing techniques that you learned about in *Chapter 4* and *Chapter 5* will be used in this chapter to build a **WAP** controller. You will investigate the operation of a Python-based Wi-Fi scanner project packaged with the UiFlow firmware, version 1.7.5. This hands-on investigation will provide context for the C++ version in which you will enhance its visual appeal. Before its visual enhancement, you will be introduced to developing C++ code on an M5Stack Core with a Hello World application.

By the end of this chapter, you will have learned how to carry out the following technical tasks with the M5Stack Core and Wi-Fi technology:

- Draw a system block diagram of an ESP32 Wi-Fi application

- Set up the M5Stack Core controller to access a wireless home network

- Develop and test a Wi-Fi scanner device using the M5Stack Core controller

- Develop and test a Wi-Fi scanner device with an LED indicator using the M5Stack Core controller

- Develop and test a Wi-Fi-based WAP and web server using an M5Stack Core controller

In this chapter, we are going to cover the following main topics:

- An introduction to Wi-Fi with the ESP32
- Wi-Fi setup for the M5Stack Core controller
- Detecting Wi-Fi networks with an M5Stack Core scanner – part 1
- Detecting Wi-Fi networks using an M5Stack Core scanner with an LED indicator – part 2
- Creating a WAP and a web server with an M5Stack Core controller

Technical requirements

To engage with the chapter's learning content, you will need the **M5GO IoT Starter kit** to explore Wi-Fi IoT technologies. You will be introduced to the M5Burner to download and install firmware onto the ESP32 microcontroller-based controller. Further, the **Arduino integrated development environment (IDE)** software with the M5Stack library will be required to build and run the M5Stack Core Wi-Fi applications.

You will require the following:

- The M5GO IoT Starter kit
- The M5Burner software
- The Arduino IDE software
- The M5Stack library
- The M5Stack Arduino API
- A Snap Circuits Bicolor LED
- Snap Circuits electrical jumper wires
- A standard four-connector-based jumper wire harness
- A smartphone

Here is the GitHub repository for the software resources: `https://github.com/PacktPublishing/M5Stack-Electronic-Blueprints/tree/main/Chapter08`

An introduction to Wi-Fi with the ESP32

The M5Stack Core ESP32 microcontroller's Wi-Fi system architecture consists of two subsystems: **Wi-Fi media access control (MAC)** and the **baseband**. The Wi-Fi **MAC** address is a unique identifier that is assigned to a **network interface controller (NIC)**. The NIC allows us to address the network used in the Wi-Fi-based communication system. The MAC address for the M5Stack Core controller is shown in *Figure 8.1*:

```
Chip is ESP32D0WDQ6 (revision 1)
Features: WiFi, BT, Dual Core, 240MHz, VRef calibration in efuse
MAC: b8:f0:09:c6:16:c4
```

Figure 8.1 – The M5Stack Core MAC address

The Wi-Fi baseband in a communication system is the **modulating signal** or **intelligence** sent through a single channel. The intelligence or digital data stream is sent through as the information through a single channel media. Wi-Fi baseband communication uses bidirectional communication transmitted through the single channel media to send and receive digital data. *Figure 8.2* illustrates the ESP32 communication system block diagram.

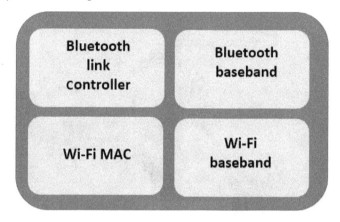

Figure 8.2 – A block diagram for the ESP32 communication system

As observed, the ESP32 Wi-Fi subsystem blocks align with the **Bluetooth link controller** and the **Bluetooth baseband**. In general, the ESP Wi-Fi subsystem blocks implement **transmission control protocol/internet protocol** (**TCP/IP**) and the full **802.11 b/g/n Wi-Fi MAC** protocols. TCP/IP is a seven-layer model that describes the ESP32 microcontrollers' interaction with a communication network. Therefore, TCP/IP describes the function of the 802.11 b/g/n Wi-Fi MAC protocols. You now know about ESP32 Wi-Fi communication. In the next section, you will learn how to set up the M5Stack Core's Wi-Fi.

Quiz 1

_____ or digital data stream is sent through as information.

Wi-Fi setup for the M5Stack Core controller

In this section, you will learn how to set up Wi-Fi for the M5Stack Core controller by adding the appropriate libraries to the Arduino IDE. The appropriate libraries are the M5Stack Core C++ files.

These files will have the coding resources to allow your M5Stack Core controller to connect with a home Wi-Fi network. To obtain access to M5Stack Core libraries, read the installation instructions at the following website: `https://docs.m5stack.com/en/quick_start/m5core/arduino`. On this website, you will find instructions on installing the Arduino IDE with M5Stack Core libraries. Be sure to include the M5Stack Core 2 libraries during the selection and installation of the C++ files. With the installation complete, you will find a variety of coding projects and examples to explore within the Arduino IDE using the M5Stack Core. The projects and examples of interest will be the Wi-Fi projects. *Figure 8.3* presents the M5Stack Core Wi-Fi coding projects and examples:

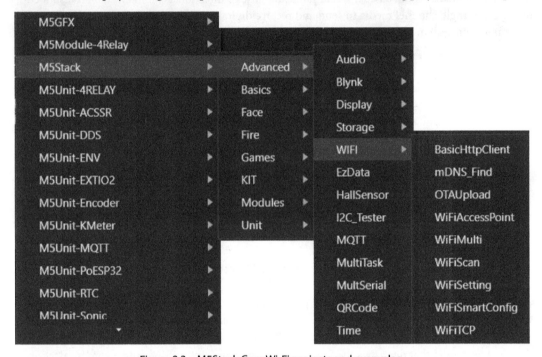

Figure 8.3 – M5Stack Core Wi-Fi projects and examples

After completing the installation steps presented in the Arduino M5Stack Core online article, you will perform a simple coding exercise. This coding exercise is a mini tutorial on programming the M5Stack Core using the C++ language. Therefore, you will go to the Arduino website to obtain the Arduino API information for the M5Stack Core. The Arduino API website is located here: `https://docs.m5stack.com/en/quick_start/m5core/arduino`.

The mini-C++ tutorial lesson consists of programming the M5Stack Core to display **Hello World** on the **TFT LCD**. As seen, an explanation of each line of code is provided for the M5Stack Core TFT LCD operation. *Figure 8.4* is the Hello World mini tutorial lesson:

```
#include <M5Stack.h>

/* After M5Core is started or reset
the program in the setUp () function will be run, and this part will only be run once.
After M5Core is started or reset, it will start to execute the program in the setup() function, and this part will only be
executed once. */
void setup(){
  M5.begin(); //Init M5Core. Initialize M5Core
  M5.Power.begin(); //Init Power module. Initialize the power module
                   /* Power chip connected to gpio21, gpio22, I2C device
                      Set battery charging voltage and current
                      If used battery, please call this function in your project */
  M5.Lcd.print("Hello World"); // Print text on the screen (string) Print text on the screen (string)
}

/* After the program in setup() runs, it runs the program in loop()
The loop() function is an infinite loop in which the program runs repeatedly
After the program in the setup() function is executed, the program in the loop() function will be executed
The loop() function is an endless loop, in which the program will continue to run repeatedly */
void loop() {

}
```

Figure 8.4 – The C++ Hello World code

Open the Arduino IDE and create a new file called `Hello World`. Create a folder on your hard drive with the same name. Save the C++ application in the folder just created. Next, begin typing the code into the editor. For convenience and ease, you can copy the code from the Arduino IDE M5Stack Core web page. With your mouse, scroll down to you see the section titled `Hello World`. Before uploading the code to the M5Stack Core, you can see the code structure. If you are familiar with the Arduino IDE, the programming structure is the same except for the `M5.object.function` instruction.

This new programming instruction is from the M5Core API library. The library is imported into the Arduino code using the `# include <M5Stack.h>` programming header. The programming header provides all the physical features of the M5Stack, which include the GPIO, speaker, system, RGB LED, buttons, and IMU. To initialize or start the M5Stack Core power module, the `M5.begin()` and `M5.Power.begin()` programming instructions are needed to ensure the M5Stack Core will operate properly. The `M5.Lcd.print` instruction will display **Hello World** on the TFT LCD. To ensure the text can be seen from a reasonable distance on the M5Stack Core TFT LCD, the `M5.Lcd.setTextSize(2)` programming instruction is used. You will include this coding instruction below the `M5.Lcd.print` code. The line number will be *13* for inserting the new programming instruction. *Figure 8.5* illustrates the location of the new code:

```
7   void setup(){
8     M5.begin(); //Init M5Core. Initialize M5Core
9     M5.Power.begin(); //Init Power module. Initialize the power module
10                    /* Power chip connected to gpio21, gpio22, I2C device
11                       Set battery charging voltage and current
12                       If used battery, please call this function in your project */
13    M5.Lcd.setTextSize(2);
14    M5.Lcd.print("Hello World"); // Print text on the screen (string) Print text on the screen (string)
15  }
```

Figure 8.5 – The location of the M5.Lcd.setTextSize(2) code instruction

To upload the `Hello World` code to the M5Stack, you will select the M5Stack as the controller board and its COM port. The uploading step is the same as for adding application code to the Arduino Uno. The selection of the M5Stack Core and the COM port is illustrated in *Figure 8.6*.

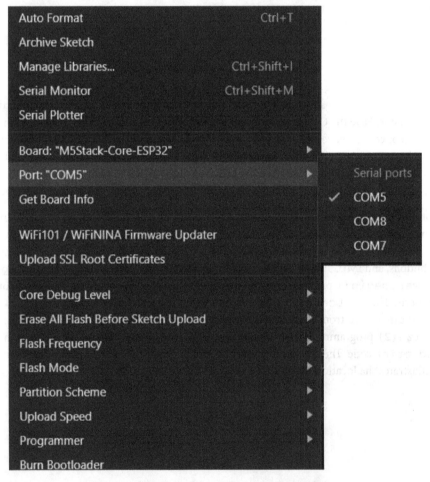

Figure 8.6 – Selection of the M5Stack Core and COM port

Interactive quiz 1

Is the **Hello World** text visible when using the `M5.Lcd.setTextSize(4)` instruction?

The final step is to upload the `Hello World` code to the M5Stack Core controller. Click on the **Upload** button identified by the arrow on the taskbar. The compilation process will begin and end by displaying the text on the M5Stack Core's TFT LCD. *Figure 8.7* illustrates **Hello World** being displayed on the M5Stack Core's TFT LCD.

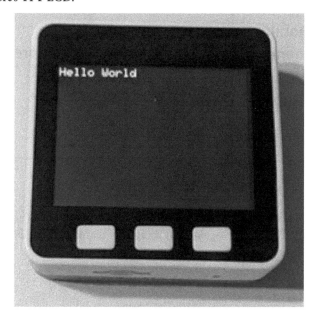

Figure 8.7 – Hello World

Congratulations on completing the mini C++ tutorial. You now have the knowledge and the skill to use the Arduino IDE to install a C++ application onto the M5Stack Core controller. The knowledge and skills learned will allow you to upload a mobile-based app to configure the Wi-Fi on the M5Stack controller.

With your mobile phone, you will install the EspTouch app. The EspTouch app allows you to configure and attach a home network to the ESP32. This mobile phone app will allow the M5Stack Core controller to detect your home network. The home network contents, such as the **IP** and **Received Signal Strength Indicator** (**RSSI**), will be displayed on the M5Stack Core controller's TFT LCD. The diagram provided in *Figure 8.8* illustrates this wireless signal detection scheme.

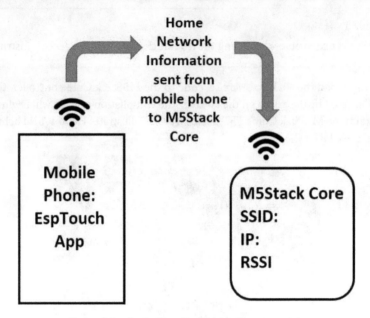

Figure 8.8 – A wireless signal detection approach

With this setup scheme, a portable remote Wi-Fi network monitor device can be realized. To obtain the EspTouch app (*the app logo is a red-colored WiFi symbol and in iOS, the app name is Espressif Esptouch*), go to the Apple store or the Google Play store. Once the app has been found, select it from the list of apps, and install it on your mobile phone. Open the EspTouch app to display the start screen. You will see **EspTouch** and **EspTouch V2** on the screen: select **EspTouch**:

Figure 8.9 – EspTouch app home screen

After selecting **EspTouch**, the following screen shown in *Figure 8.10* will be presented.

Note

RSSI is a measurement of the wireless device's capability to receive a signal from an access point or router.

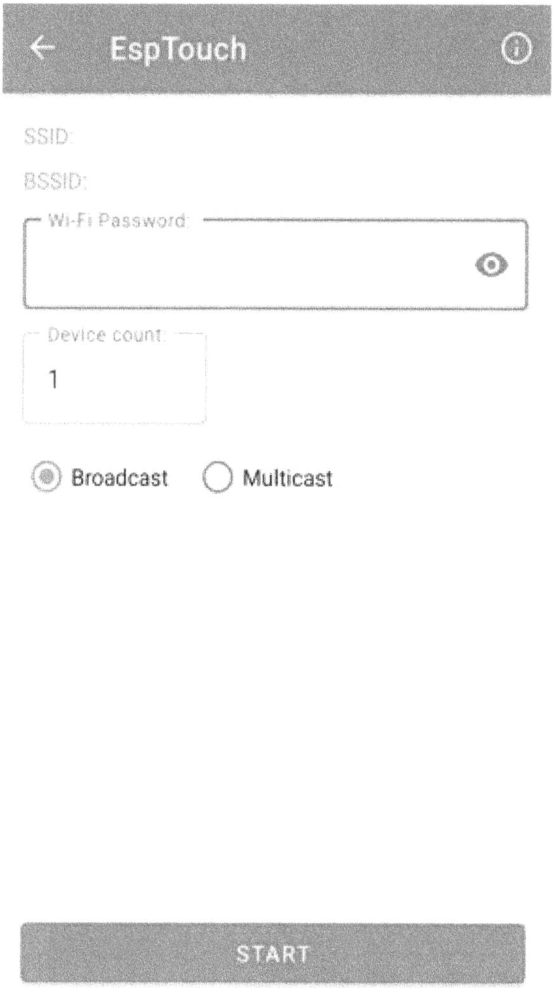

Figure 8.10 – The EspTouch Start screen

You will see your **Service Set Identifier (SSID)** number and the **Basic Service Set Identifier (BSSID)** of your home network. With the app open on your mobile phone, you will proceed to upload the `WiFiSmartConfig.ino` software onto the M5Stack Core. You will find this software by selecting **File | Examples | M5Stack | Advanced | WiFi | WiFiSmartConfig** within the Arduino IDE. *Figure 8.11* illustrates the selection steps presented.

> **Note**
>
> SSID is the name of the Wi-Fi network. The BSSID is the MAC physical address of the access point or wireless router that is used by the network to connect to the Wi-Fi.

Figure 8.11 – Selecting the WiFiSmartConfig.ino application

You can upload the `WiFiSmartConfig.ino` code using the previous steps illustrated in *Figure 8.6* to the M5Stack Core. *Figure 8.12* shows the partial code for the `WiFiSmartConfig.ino` wireless application. The original text font size is *1*. Add the `M5.Lcd.setTextSize(2)` instruction before each `M5.Lcd.print()` line of code to make the text readable on the **TFT LCD**. *Figure 8.13* illustrates this code modification step. This step will allow some experimentation with the UI layout design to align with readability. The exploration with the `M5.Lcd.setTextSize(2)` instruction will demonstrate the importance of creating UI layouts that make text readable. Making code improvements will allow you to experiment with the TFT LCD using C++ code:

```
18    #include <M5Stack.h>
19
20    #include "WiFi.h"
21
22    void setup() {
23        M5.begin();                        // Init M5Core. 初始化 M5Core
24        M5.Power.begin();                  // Init power   初始化电源模块
25        WiFi.mode(WIFI_AP_STA);            // Set the wifi mode to the mode compatible with
26                                           // the AP and Station, and start intelligent
27                                           // network configuration
28        WiFi.beginSmartConfig();           // 设置wifi模式为AP 与 Station
29                                           // 兼容模式,并开始智能配网
30
31        // Wait for the M5Core to receive network information from the phone
32        //等待M5Core接收到来自手机的配网信息
33        M5.Lcd.println(
34            "\nWaiting for Phone SmartConfig.");  // Screen print format string.
35                                                  // 屏幕打印格式化字符串
36        while (!WiFi.smartConfigDone()) {  // If the smart network is not completed.
37                                           // 若智能配网没有完成
38            delay(500);
39            M5.Lcd.setTextSize(2);
```

Figure 8.12 – Partial WiFiSmartConfig.ino application code

```
38            delay(500);
39            M5.Lcd.setTextSize(2);
40            M5.Lcd.print(".");
41        }
42        M5.Lcd.setTextSize(2);
43        M5.Lcd.println("\nSmartConfig received.");
44        M5.Lcd.setTextSize(2);
45        M5.Lcd.println("Waiting for WiFi");
46        while (
47            WiFi.status() !=
48            WL_CONNECTED) {  // M5Core will connect automatically upon receipt of
49                             // the configuration information, and return true if
50                             // the connection is successful.
51                             // 收到配网信息后M5Core将自动连接, 若连接成功将返回true
52            delay(500);
53            M5.Lcd.setTextSize(2);
54            M5.Lcd.print(".");
55        }
```

Figure 8.13 – Improving the text font size with M5.Lcd.setTextSize(2)

You will tap the device count box within the EspTouch app. Enter 1 into the box and check the checkbox on your mobile view. Touch the **Start** button at the bottom of the EspTouch app. Touch **ok** when the EspTouch app displays the BSSID number. Press the power button on the M5Stack Core. Within a few seconds, you should see your home network's IP and the RSSI information displayed on the M5Stack Core. *Figure 8.14* illustrates the completed step. Congratulation on capturing your Wi-Fi setup information using the M5Stack Core. In this next section, you will continue your Wi-Fi investigation by building a handheld wireless scanner.

Figure 8.14 – Home network information displayed on the M5Stack Core

> **Note**
> Make sure the M5Stack is in internet mode and on the same WiFi as the phone or tablet.

Detecting Wi-Fi networks with an M5Stack scanner – part 1

You have seen that the M5Stack Core has a Wi-Fi subsystem consisting of a baseband and MAC electronic circuits. To aid these hardware elements, the software allows the configuration of these circuits with the ESP32 microcontroller core subsystem. The M5Stack Core will allow us to assess a home Wi-Fi network displayed on a 2-inch TFT LCD. The application detection code will help the M5Stack Core cohesively stitch the Wi-Fi circuits to the ESP32 core subsystem. This is all achieved using the Wi-Fi scan code found within the Arduino IDE examples. You will use the M5Stack Core example **WiFiScan** code to assess and display the wireless networks within your home. To access the C++code, select **File** | **Examples** | **M5Stack** | **Advanced** | **WiFi** | **WiFiScan** with your mouse. *Figure 8.15* shows the selection sequence for obtaining the **WiFiScan** code:

Figure 8.15 – Obtaining the WiFiScan code

Upon opening the **WiFiScan** code within the Arduino IDE, you will notice the default font size within the software listing. To improve the readability of the fonts, the M5.Lcd.setTextSize(2) instruction will be used. *Figure 8.16* illustrates the partially improved **WiFiScan** code.

Interactive quiz 2

What is the maximum acceptable font size to use for improving the readability of the SSID and RSSI data?

The usability of the M5Stack Core controller will be improved by making the font size readable. The default text size requires an element of magnification so that the text can be read. Again, the M5.Lcd. setTextSize() instruction will improve this usability concern:

```
15   #include "WiFi.h"
16
17   void setup() {
18       M5.begin();              // Init M5Stack.  初始化M5Stack
19       M5.Power.begin();        // Init power  初始化电源模块
20       WiFi.mode(WIFI_STA);     // Set WiFi to station mode and disconnect from an AP
21                                // if it was previously connected.
22                                // 将WiFi设置为站模式，如果之前连接过AP，则断开连接
23       WiFi.disconnect();       // Turn off all wifi connections.  关闭所有wifi连接
24       delay(100);              // 100 ms delay.  延迟100ms
25       M5.Lcd.setTextSize(2);
26       M5.Lcd.print("WIFI SCAN");  // Screen print string.  屏幕打印字符串
27   }
28
29   void loop() {
30       M5.Lcd.setCursor(0, 0);  // Set the cursor at (0,0).  将光标设置在(0,0)处
31       M5.Lcd.println("Please press Btn.A to (re)scan");
32       M5.update();  // Check the status of the key.  检测按键的状态
33       if (M5.BtnA.isPressed()) {  // If button A is pressed.  如果按键A按下
34           M5.Lcd.clear();              // Clear the screen.  清空屏幕
35           M5.Lcd.setTextSize(2);
36           M5.Lcd.println("scan start");
```

Figure 8.16 – Improved readability for WiFiScan

After the modifications have been made to the **WiFiScan** C++ code, you can upload the software to the M5Stack Core. The code will be compiled and then uploaded to the M5Stack Core's ESP32 microcontroller. Once the code is uploaded to the M5Stack Core, the **Please press Btn.A to(re)scan** message will be displayed on the TFT LCD. *Figure 8.17* shows the text displayed on the TFT LCD.

Figure 8.17 – The WiFiScan code running on the M5Stack Core

When you press **Button A** to start the scan, the Wi-Fi networks within your home will be displayed on the M5Stack Core's TFT LCD. The RSSI data for each network will be displayed as well. You can repeat the scan by pressing **Button A**. Congratulations – you have successfully programmed the M5Stack Core to scan and display wireless networks in your home.

Another version of a Wi-Fi scanner that you can evaluate through experimentation is within the M5Stack Core's **firmware**. To obtain the Wi-Fi scanner, the M5Burner software must be used. The M5Burner software can be obtained from here: `https://docs.m5stack.com/en/download`. The M5Burner software is available for Windows, macOS, and Linux **operating systems** (**OSes**). *Figure 8.18* shows the available M5Burner software for the listed OSes.

| UIFLOW FIRMWARE BURNING TOOL

NO	Name	Download
1	M5Burner Win10 x64 v3.0	↓
2	M5Burner MacOS x64 v3.0	↓
3	M5Burner Linux x64 v3.0	↓

Figure 8.18 – The M5Burner firmware software

You will download the appropriate software for a desktop **personal computer** (**PC**) or laptop computer OS. Install it and open the M5Burner firmware software. With the software open, click the down arrow and select **v1.7.5** from the drop-down list box located on the right. Click on the **Burn** button to start the installation of the 1.7.5 firmware version onto the M5Stack Core. *Figure 8.19* illustrates the M5Burner firmware software. The M5Burner firmware software, once installed on the M5Stack Core, will reset the device automatically.

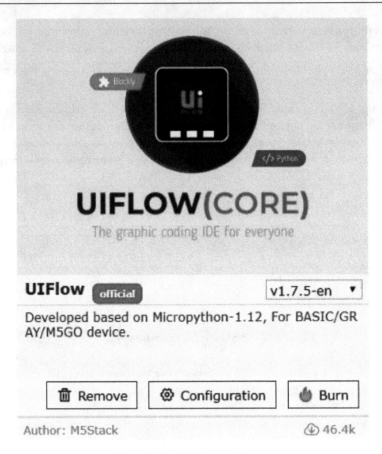

Figure 8.19 – M5Burner software

Interactive quiz 2

Using *Figure 8.16*, modify the **WiFiScan** code to start the scan with the **Button B** input control.

Once the firmware has been installed, the UiFlow splash screen will appear on the TFT LCD. To obtain the Wi-Fi scanner application, press **Button A** (the **APP** icon) to view the list of MicroPython applications. A partial list of MicroPython applications is shown in *Figure 8.20*.

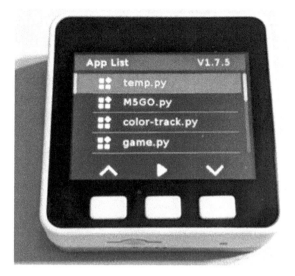

Figure 8.20 – M5GO MicroPython applications

You will use the down arrow located above **Button C** to scroll to the `wifi_scaner.py` application:

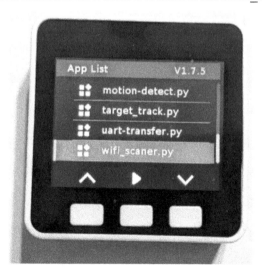

Figure 8.21 – M5GO MicroPython wifi_scaner application

Select the application by pressing **Button B** on the M5Stack Core. Upon making this selection, the Wi-Fi scanner will immediately list the wireless networks within your home. The scan will continue to repeat the search and display function. You can stop this wireless scan operation by pressing the **ON** power button once. This step will turn off the M5Stack Core controller, thus allowing the startup of the Wi-Fi scanner. The next stage in the Wi-Fi scanner development is to provide visual feedback.

Detecting Wi-Fi networks with an M5Stack scanner with an LED indicator – part 2

You have investigated two Wi-Fi scanner designs using the M5Stack Core controller. The first Wi-Fi scanner design allowed you to manipulate the C++ code to change the text font size. The second Wi-Fi scanner design provided a quick method to assess your wireless home networks. Having gone through these two design approaches, you now have the knowledge and technical skills to enhance the Wi-Fi scanner. The enhancement will require adding an LED indicator. To make this physical visual feature improvement to the Wi-Fi scanner, a software modification is needed. The concept for the Wi-Fi scanner with an LED indicator is shown in *Figure 8.22*:

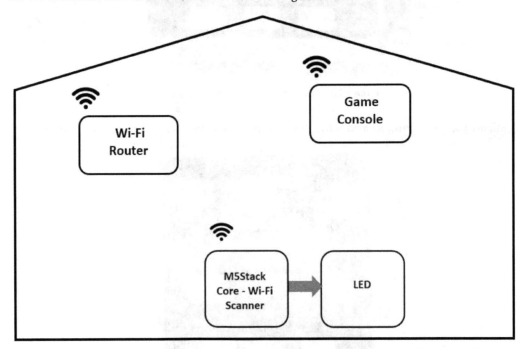

Figure 8.22 – A Wi-Fi scanner with an LED indicator

The diagram illustrates a home network in which the LED provides a visual indication of detecting wireless devices. The M5Stack Core Wi-Fi scanner that we investigated in the first activity provided IP and RSSI information on the TFT LCD. The electronic scanner seamlessly connects to the home network using a Wi-Fi baseband subcircuit. The LED provides a visual indication of the Wi-Fi scanner completing its scan of wireless devices detected within a home network. Upon starting a new scan, the LED will be off.

Therefore, this project is developed with two parts: hardware and software. The hardware development consists of wiring an LED circuit to the M5Stack Core controller. Port B has a digital output pin (**G26**) that allows you to connect external circuits for control. You will wire the LED circuit to this digital output pin to provide a visual indication of the completed scan of wireless devices connected within a home network. You will use the electronic circuit schematic diagram in *Figure 8.23* to wire the LED circuit to port B's **G26** digital output pin.

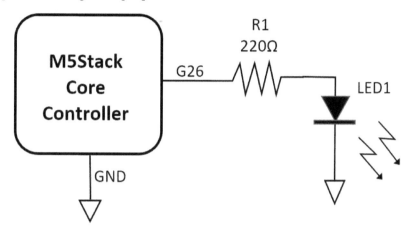

Figure 8.23 – Electronic circuit schematic diagram for a Wi-Fi scanner with an LED indicator

Although *Figure 8.23* shows discrete electronic components wired in series to a digital pin, **G26**, a Snap Circuits LED may be used. The Snap Circuits LED component, **D10**, is a bicolor component. Reversing the voltage source across the bicolor LED provides two unique colors. The Snap Circuits LED component will emit a red- or yellow-colored light based on the reversal of an applied voltage source. Therefore, the output color indication of the scanned Wi-Fi network-connected devices can be changed by the reversal of port B's **G26** digital pin and the electrical ground. *Figure 8.24* shows the two electronic circuit schematic diagrams for emitting red or yellow light from the bicolor LED:

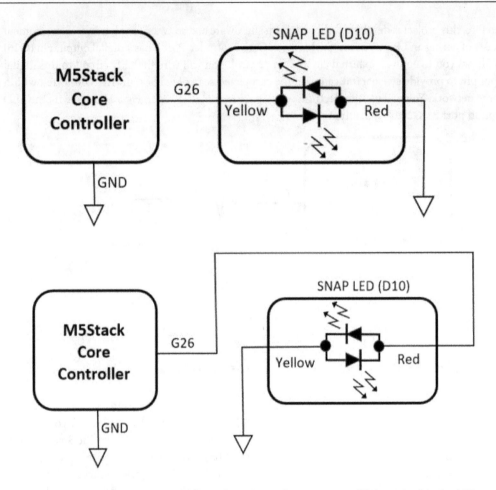

Figure 8.24 – Electronic circuit schematics for red or yellow emission of light with a bicolor LED

You will use the four-wire electrical jumper harness attached to the Snap Circuits bicolor LED component. Using the Snap connectors, you can make reverse the wires between the bicolor LED and the M5Stack Core controller conveniently and easily. *Figure 8.25* illustrates a four-wire electrical jumper harness attachment between the M5Stack Core controller and the Snap Circuits bicolor LED component.

Quiz 2

Reviewing the two electronic circuit schematic diagrams shown in *Figure 8.24*, which circuit will allow the bicolor LED to emit red light via the M5Stack Core controller?

a) The top circuit

b) The bottom circuit

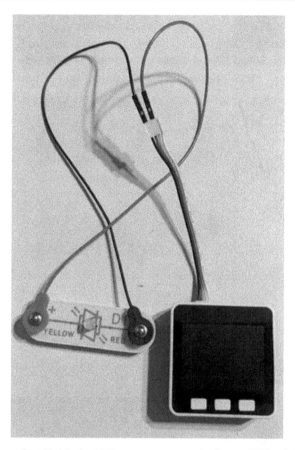

Figure 8.25 – Snap Circuits bicolor LED component attached to an M5Stack Core controller

With the two electronic circuit schematic diagrams presented, you will be able to test the bicolor LED of the Snap Circuits component quite easily. The hardware portion of this Wi-Fi scanner modification activity is completed. The final step requires the inclusion of digital output port B pin, **G26**, to operate the bicolor LED in the Wi-Fi scanner C++ code. The C++ code modification will use the traditional `digitalWrite()` and `pinMode()` Arduino Uno programming instructions.

Interactive quiz 3

With the two electronic circuit schematic diagrams presented in *Figure 8.24*, is it possible to provide an audible alarm status indicator with a piezo buzzer?

The first modification to the Wi-Fi code is setting **G26** as the digital output pin of port B. To accomplish this coding task, the `pinMode ( )` instruction is used. The `void setup ( )` function is the part of the C++ code in question. *Figure 8.26* shows the inclusion of the `pinMode ( )` instruction when establishing **G26** as a digital `OUTPUT` pin within the `void setup ( )` function:

```
17   void setup() {
18       M5.begin();              // Init M5Stack.   初始化M5Stack
19       M5.Power.begin();        // Init power   初始化电源模块
20       WiFi.mode(WIFI_STA);     // Set WiFi to station mode and disconnect from an AP
21                                // if it was previously connected.
22                                // 将WiFi设置为站模式，如果之前连接过AP，则断开连接
23       WiFi.disconnect();       // Turn off all wifi connections.   关闭所有wifi连接
24       delay(100);              // 100 ms delay.   延迟100ms
25       M5.Lcd.setTextSize(2);
26       M5.Lcd.print("WIFI SCAN");  // Screen print string.   屏幕打印字符串
27       pinMode(26, OUTPUT);
28   }
```

Figure 8.26 – Defining pin G26 as OUTPUT

With port B set up as a digital `OUTPUT` pin, you use it to turn on the Snap Circuits bicolor LED component. The Arduino C++ code instruction to use is `digitalWrite ( )`. The bicolor LED will turn on after the complete scan of the wireless network and display them on M5Stack Core's TFT LCD. The line number at which to include the `digitalWrite(26,1)` instruction is shown next.

```
51               M5.Lcd.setTextSize(2);
52               M5.Lcd.printf("%d:", i + 1);
53               M5.Lcd.setTextSize(2);
54               M5.Lcd.print(WiFi.SSID(i));
55               M5.Lcd.setTextSize(2);
56               M5.Lcd.printf("(%d)", WiFi.RSSI(i));
57               M5.Lcd.setTextSize(2);
58               M5.Lcd.println(
59                   (WiFi.encryptionType(i) == WIFI_AUTH_OPEN) ? " " : "*");
60               digitalWrite(26, 1);
```

Figure 8.27 – Code location to turn on the G26 digital pin

The bicolor LED will turn off when **Button A** on the TFT LCD is pressed. The bicolor LED will also turn off when there are no discovered scanned Wi-Fi devices or networks. *Figure 8.28* shows the location of the `digitalWrite(26,0)` instruction to include in the Wi-Fi scanner code:

```
30    void loop() {
31        M5.Lcd.setCursor(0, 0);  // Set the cursor at (0,0). 将光标设置在(0,0)处
32        M5.Lcd.println("Please press Btn.A to (re)scan");
33        M5.update();  // Check the status of the key. 检测按键的状态
34        if (M5.BtnA.isPressed()) {  // If button A is pressed. 如果按键A按下
35            M5.Lcd.clear();              // Clear the screen. 清空屏幕
36            M5.Lcd.setTextSize(2);
37            M5.Lcd.println("scan start");
38            digitalWrite(26, 0);
```

```
41              if (n == 0) {  // If no network is found. 如果没有找到网络
42                  M5.Lcd.setTextSize(2);
43                  M5.Lcd.println("no networks found");
44                  digitalWrite(26, 0);
```

Figure 8.28 – Turning off the bicolor LED with the digitalWrite(26, 0) instruction

As captured in the C++ code snippets, the value of 2 establishes the text size for the Wi-Fi scanner application. The M5.Lcd.setTextSize( ) instruction has a numerical range of 0 to 7. As a UI layout design usability experiment, try finding the optimal text size by changing the parameter's value. Record the value to use in future M5Stack Core controller projects. Further, replace the Snap Circuits bicolor LED component with an RGB unit to give a new look to detecting the wireless devices and networks in the home. You will now complete the final project of this chapter: creating a WAP and a **web server** with an M5Stack Core controller.

Quiz 3

If the following lines of code are provided for an external wired LED during a completed wirelessly network scan, what is the visual effect?

```
digitalWrite(26,1);

delay(500);

delayWrite(26,0);

delay(500);
```

Creating an access point and web server with an M5Stack Core controller

In this final chapter project, you will build an access point and web server using an M5Stack Core controller and a smartphone. You will use the M5Stack Core to create an access point to connect and control a wireless device. The WAP will be accessible with an SSID and URL network address. The smartphone will allow access to the wireless network by connecting to the M5Stack Core 2 WAP SSID. The result is connecting to a web server that allows you to control an external device wired to an M5Stack Core controller. The concept for the WAP and **web-server-based controller** is shown in *Figure 8.29*.

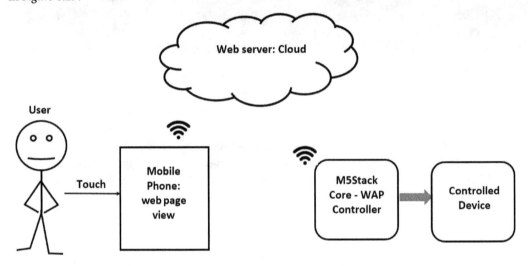

Figure 8.29 – Use case model concept for the WAP web server controller

The user will touch the mobile phone screen to interact with the web page UI controls. The **web server** web page UI controls allow a simple **ON/OFF** control capability. The M5Stack Core WAP controller receives the ON/OFF commands from the web server web page based on the network established from the web server and the M5Stack Core WAP controller. The M5Stack Core WAP controller will turn the wire-controlled device on and off. Small electromechanical loads such as motors and solenoids, LEDs, transistor relay circuits, and piezo buzzers are examples of controlled devices. The use case model presented in *Figure 8.29* can be modified to show an electronic circuit schematic diagram of the M5Stack Core WAP controller. *Figure 8.30* illustrates the new WAP web server controller diagram. In this diagram, additional details have been added as they relate to the controlled device circuit components. This basic electronic circuit diagram presents the concept of the M5Stack Core as a practical and functional controller capable of operating a variety of discrete visual, audible, and **electromechanical actuator** loads. The knowledge and skills obtained in previous chapters exploring interfacing concepts using littleBits electronic modules, Snap Circuits, and Arduino will be applied in the operational scheme of this controlled device.

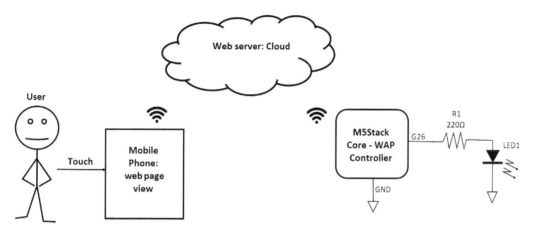

Figure 8.30 – Detailed WAP web server controller diagram

The M5Stack Core WAP controller is the electronic circuit schematic diagram illustrated in *Figure 8.24* to provide a completed Wi-Fi network scan. To explore the configuration of this WAP controller's wiring method, *Figure 8.31* expands on port B and its alignment with the G26 output and attachment to the M5Stack Core. Another important element regarding port B and the jumper harness is the identification of the G26 output pin:

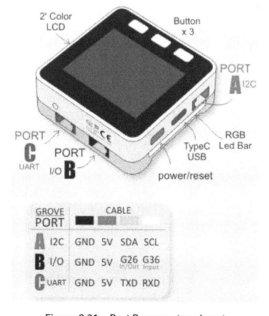

Figure 8.31 – Port B connector pinout

As noted in *Figure 8.31*, the colored wire associated with port B's **G26** digital pin is yellow. The key identifier of the **G26** pin is its physical location on the male connector. **G26** is in the third position within the male connector's construction. With this information, the color of the wire is not the true indicator of the digital control signal presence but the pin location. The four-conductor jumper wire harness color scheme regarding yellow and white can be switched. *Figure 8.32* illustrates this important fact of the reversal of color wire positioning among the two four-conductor jumper wire harness variants:

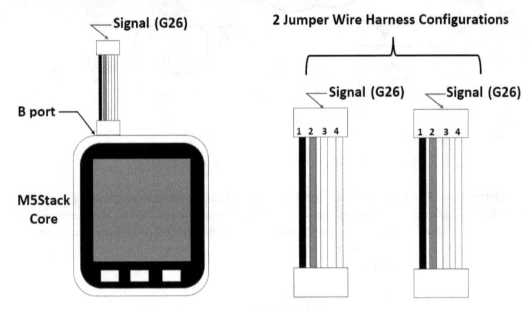

Figure 8.32 – Location of the G26 digital pin

As illustrated in *Figure 8.32*, between the two jumper wire harness configurations, what is consistent is the location of the **G26** digital pin. You now have a further understanding of the alignment of port B and the physical location of the **G26** digital pin. Another design practice to consider as it relates to the electrical wiring of specialized electronic circuits is the development of the **system block diagram**.

The system block diagram is an important design tool when developing controlled devices for the M5Stack Core WAP controller. As illustrated in *Figure 8.29* and *Figure 8.30*, the system block diagram was created with a two-level design approach. The first-level system block diagram, as presented, is conceptual, meaning very few details of specific component values and symbols are not provided. The second level system block diagram provides information on electrical and electronic component names and I/O port designators. Occasionally, electrical and electronic component names with symbol representation are provided in a second-level system block diagram. With the system block diagram you designed, the electrical-electronic interface circuits may be realized in a seamless matter.

You may now proceed to install the Wi-Fi AP software for the M5Stack Core WAP controller. The baseline C++ code to be used with the M5Stack Core controller is the `WiFiAccessPoint.ino` file. You will find this code by going to the following directory of **File | Examples | M5Stack | Advanced | WiFi | WiFiAccessPoint**. *Figure 8.33* shows where to access the `WiFiAccessPoint.ino` file. You will use the Arduino IDE to access and open this code:

Figure 8.33 – Accessing the WiFiAccessPoint code

With the `WiFiAccessPoint.ino` file, you will use the code to build a framework for the M5Stack Core controller as an IoT development platform. You will improve the code by changing the text size. In the previous project, you use the `M5.Lcd.setTextSize( )` instruction to allow the TFT LCD displayed information to be readable. *Figure 8.34* illustrates the approach of including the `M5.Lcd.setTextSize( )` instruction to improve the `WiFiAccessPoint` readability on the TFT LCD. You will add this coding instruction through the entirety of the `WiFiAccessPoint` code.

After completing this task, rename the program and save it. You now have created a software workspace in which further modifications can be carried out without damaging the original code. This approach is a good software development practice of ensuring the original code will be available as a reference to the application being developed:

```
54                    M5.Lcd.setTextSize(2);
55                    M5.lcd.print("New Client:");
56                    String currentLine =
```

Figure 8.34 – Adding the M5.Lcd.sizeText() instruction for readability

With the text size changes added to the code, upload the software to the M5Stack Core controller. Add your SSID and password to the network you would like to connect with. The location within the code is shown in *Figure 8.35*.

```
22    // Set these to your desired credentials.    设置你的热点名称和密码
23    const char *ssid     = "";
24    const char *password = "";
```

Figure 8.35 – Adding the Wi-Fi network login credentials

Attach the Snap Circuits bicolor LED to port B of the M5Stack Core controller. This task will be accomplished by using the four-wire jumper harness as the electrical connecting interface between the M5Stack Core controller's port B and the Snap Circuits bicolor LED. *Figure 8.36* illustrates the electrical connection of the controlled device (the Snap Circuits bicolor LED) to port B of the M5Stack Core controller. Turn on the M5Stack Controller by pressing the red power button located on the left side of the device.

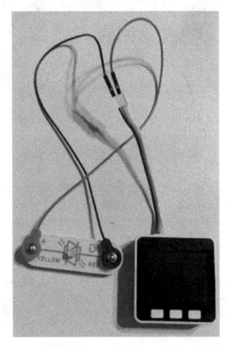

Figure 8.36 – Connecting the Snap Circuits bicolor LED to the M5Stack Core controller

You will upload the modified code to the M5Stack Core controller. The upload process consists of checking syntax errors through code compilation. If there are no errors detected during the compilation event, the upload process will display a message as shown here:

```
Output
Writing at 0x000d2753... (100 %)
Wrote 816992 bytes (521787 compressed) at 0x00010000 in 7.4 seconds (effective 881.4 kbit/s)...
Hash of data verified.

Leaving...
Hard resetting via RTS pin...
```

Figure 8.37 – Completed code compilation and upload to the M5Stack Core controller

You will see the SSID network name and the URL address to access the web server UI controls on the M5Stack Core controller TFT LCD. *Figure 8.38* illustrates this step. Congratulations on completing this project task!

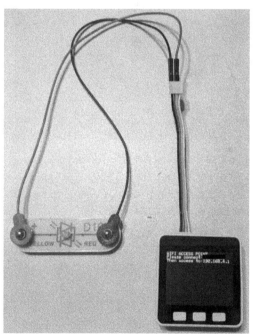

Figure 8.38 – The M5Stack Core controller connected to the UI control web server

> **Note**
> If this text size change illustrated in *Figure 8.34* does not improve the readability of the displayed information, remove it from your code and save the file.

You will connect to the network SSID used in *Figure 8.35* to obtain web server access using a smartphone. The M5Stack Core controller will provide a URL address to obtain the simple UI for operating the Snap Circuits bicolor LED with your smartphone. You will complete the final step of this project by including the `digitalWrite(26,0)` and `digitalWrite(26,1)` instructions for operating the Snap Circuits bicolor LED in the `WiFiAccessPoint` C++ code. You will use the `pinMode(26,OUTPUT)` instruction to set up the digital pin as an output driver to operate the Snap Circuits bicolor LED. *Figure 8.39* provides the line number for inserting the `pinMode(26,OUTPUT)` instruction into the code:

```
28  void setup() {
29      M5.begin();                  // Init M5Stack.  初始化M5Stack
30      M5.Power.begin();            // Init power  初始化电源模块
31      M5.lcd.setTextSize(2);  // Set text size to 2.  设置字号大小为2
32      M5.lcd.println(
33          "WIFI ACCESS POINT");  // Screen print string.  屏幕打印字符串.
34      M5.lcd.printf("Please connect:%s \nThen access to:", ssid);
35      WiFi.softAP(
36          ssid,
37          password);  // You can remove the password parameter if you want the AP
38                      // to be open.  如果你想建立开放式热点,可以删除密码
39      IPAddress myIP = WiFi.softAPIP();  // Get the softAP interface IP address.
40                                         // 获取AP接口IP地址
41      M5.lcd.println(myIP);
42      server.begin();  // Start the established Internet of Things network server.
43                       // 启动建立的物联网网络服务器
44      pinMode(26, OUTPUT);
45  }
```

Figure 8.39 – Defining G26 as an output digital pin

Figure 8.40 illustrates where the `digitalWrite(26,0)` and `digitalWrite(26,1)` instructions will be included in the `WiFiAccessPoint` C++ code:

```
116                     if (currentLine.endsWith("GET /High")) {
117                         M5.Lcd.print("ON\n");
118                         digitalWrite(26, 1);
119                     } else if (currentLine.endsWith("GET /Low")) {
120                         M5.Lcd.print("OFF\n");
121                         digitalWrite(26, 0);
122                     }
```

Figure 8.40 – Line number locations for inserting digitalWrite() instructions

---

Interactive quiz 3

Change the value of `M5.Lcd.setTextSize( )` to 4. Did the new parameter improve the readability of the displayed text?

After adding `digitalWrite(26,0)` and `digitalWrite(26,1)` into the modified `WiFiAccessPoint` C++ code, upload the software to the M5Stack Core controller. You will then enter the URL address displayed on the M5Stack Core to access the web server UI LED controls into a web browser on the smartphone. *Figure 8.41* illustrates the web server UI LED controls:

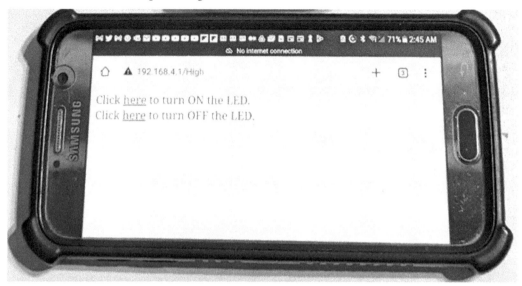

Figure 8.41 – Basic UI LED controls displayed on a smartphone

You will touch the **Click here to turn ON the LED** web server button with your finger. Upon completing this action on your smartphone, the M5Stack Core controller will turn on the Snap Circuits bicolor LED. Depending on the wiring configuration used (*see Figure 8.24*), the LED will emit red or yellow light. *Figure 8.42* illustrates the task of turning on the LED using your smartphone. If the Snap Circuits LED does not turn on, make sure the four-jumper-wire harness is properly inserted into the M5Stack Core controller's port B. Check the electrical snaps attached to the LED, along with the wires, which should be properly inserted into the jumper wire harness female connector ground and pin location 3 cavities. Finally, check to see whether you are attached to an SSID network on your smartphone. If these checks are completed, try operating the Snap Circuits LED with your smartphone.

Interactive activity

With the LED turned on, reverse the snap connections to display the other color.

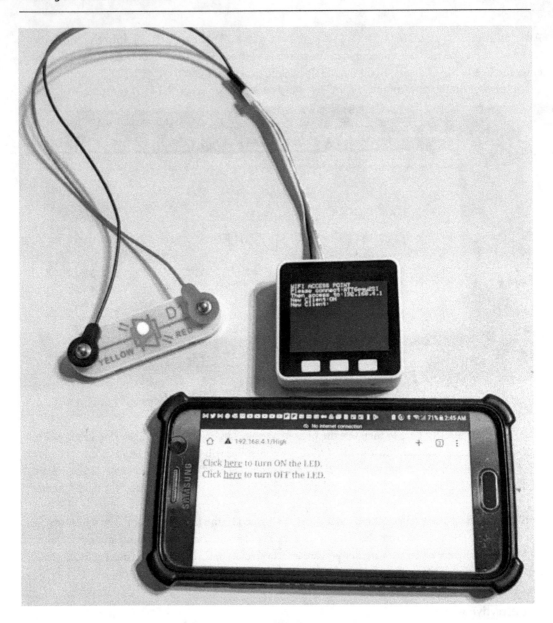

Figure 8.42 – The Snap Circuits bicolor LED operated by the WAP server

You will touch the **Click here to turn OFF the LED** web server button with your finger. Upon completing the action on your smartphone, the M5Stack Core controller will turn off the Snap Circuits bicolor LED. Depending on the wiring configuration used (*see Figure 8.24*), the LED will stop emitting either red or yellow light. *Figure 8.43* illustrates the task of turning off the LED using your smartphone.

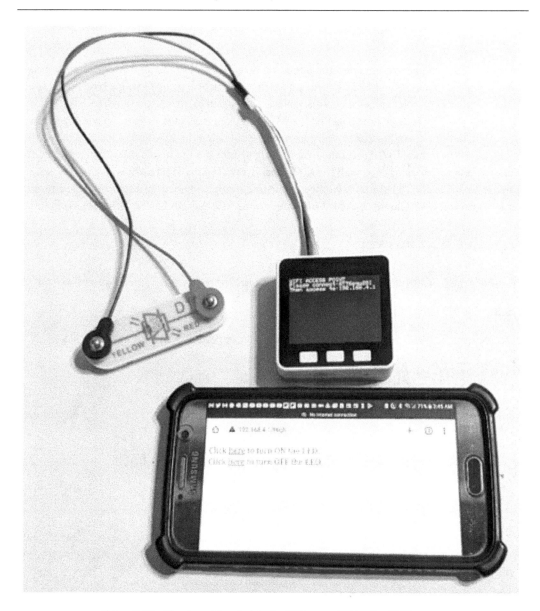

Figure 8.43 – The Snap Circuits bicolor LED operated by the WAP server

Again, if the Snap Circuits LED does not turn off, make sure the four-jumper-wire harness is properly inserted into port B of the M5Stack Core controller. Check the electrical snaps attached to the LED, along with the wires, which should be properly inserted into the jumper wire harness female connector ground and pin location 3 cavities. Finally, check to see whether you are attached to an SSID network on your smartphone. If these checks are completed, try turning off the Snap Circuits LED with your smartphone.

Summary

Congratulations – you have completed the hands-on activities and quizzes in this chapter. In this chapter, you learned about the ESP32 Wi-Fi subsystem architecture by analyzing a block diagram. With this knowledge of system block diagrams, you learned how to create one for developing an ESP32 Wi-Fi application. You learned how to set up the M5Stack Core controller to access a wireless home network. You developed and tested a Wi-Fi scanner device using the M5Stack Core controller. In analyzing a C++ code-based Wi-Fi scanner, you investigated the operation of a MicroPython-created network scanning device packaged with the M5Burner firmware software. Lastly, with the C++ skills obtained in creating a Wi-Fi scanner, you were able to develop, enhance, and test a WAP and web server to control a SNAP Circuits bicolor LED wired to the M5Stack controller with a smartphone.

Interactive quiz answers

Interactive quiz 1: Yes, the size is twice as large compared to `M5.Lcd.setTextSize(2)`.

Interactive quiz 2: `M5.BtnB.isPress( )`

Interactive quiz 3: Yes, the ON/OFF control data is quite readable from several inches away.

Quiz answers

Quiz 1: The intelligence

Quiz 2: The bottom circuit

Quiz 3: The LED will flash on and off for 1 second.

Index

Symbols

`Packt.com`

Subscribe to our online digital library for full access to over 7,000 books and videos, as well as industry leading tools to help you plan your personal development and advance your career. For more information, please visit our website.

Why subscribe?

- Spend less time learning and more time coding with practical eBooks and Videos from over 4,000 industry professionals

- Improve your learning with Skill Plans built especially for you

- Get a free eBook or video every month

- Fully searchable for easy access to vital information

- Copy and paste, print, and bookmark content

Did you know that Packt offers eBook versions of every book published, with PDF and ePub files available? You can upgrade to the eBook version at `packt.com` and as a print book customer, you are entitled to a discount on the eBook copy. Get in touch with us at `customercare@packtpub.com` for more details.

At `www.packt.com`, you can also read a collection of free technical articles, sign up for a range of free newsletters, and receive exclusive discounts and offers on Packt books and eBooks.

Other Books You May Enjoy

If you enjoyed this book, you may be interested in these other books by Packt:

The Ultimate Guide to Informed Wearable Technology

Christine Farion

ISBN: 9781803230597

- Construct sewable electronic circuits with conductive thread and materials
- Discover the features of LilyPad, Gemma, Circuit Playground, and other boards
- Use various components for listening, moving, sensing actions, and visualizing outputs
- Control ESP32 development boards for IoT exploration
- Understand why and how to prototype to create interactive wearables
- Get skilled in sewing and soldering sensors to Arduino-based circuits
- Design and build a hyper-body wearable that senses and reacts
- Master a Design Innovation approach for creating wearables with a purpose

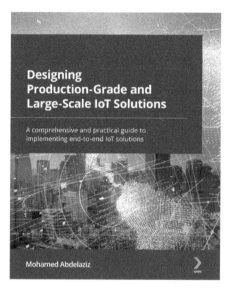

Designing Production-Grade and Large-Scale IoT Solutions

Mohamed Abdelaziz

ISBN: 9781838829254

- Understand the detailed anatomy of IoT solutions and explore their building blocks
- Explore IoT connectivity options and protocols used in designing IoT solutions
- Understand the value of IoT platforms in building IoT solutions
- Explore real-time operating systems used in microcontrollers
- Automate device administration tasks with IoT device management
- Master different architecture paradigms and decisions in IoT solutions
- Build and gain insights from IoT analytics solutions
- Get an overview of IoT solution operational excellence pillars

Packt is searching for authors like you

If you're interested in becoming an author for Packt, please visit authors.packtpub.com and apply today. We have worked with thousands of developers and tech professionals, just like you, to help them share their insight with the global tech community. You can make a general application, apply for a specific hot topic that we are recruiting an author for, or submit your own idea.

Share Your Thoughts

Now you've finished *M5Stack Electronic Blueprints*, we'd love to hear your thoughts! Scan the QR code below to go straight to the Amazon review page for this book and share your feedback or leave a review on the site that you purchased it from.

https://packt.link/r/1803230304

Your review is important to us and the tech community and will help us make sure we're delivering excellent quality content.

Download a free PDF copy of this book

Thanks for purchasing this book!

Do you like to read on the go but are unable to carry your print books everywhere?

Is your eBook purchase not compatible with the device of your choice?

Don't worry, now with every Packt book you get a DRM-free PDF version of that book at no cost.

Read anywhere, any place, on any device. Search, copy, and paste code from your favorite technical books directly into your application.

The perks don't stop there, you can get exclusive access to discounts, newsletters, and great free content in your inbox daily

Follow these simple steps to get the benefits:

1. Scan the QR code or visit the link below

https://packt.link/free-ebook/9781803230306

2. Submit your proof of purchase
3. That's it! We'll send your free PDF and other benefits to your email directly

www.ingramcontent.com/pod-product-compliance
Lightning Source LLC
Chambersburg PA
CBHW060522060326
40690CB00017B/3358

* 9 7 8 1 8 0 3 2 3 0 3 0 6 *